D1148841

This book is for
Arthur and Dusty,
Maisie, Josie, Poppy,
Sonny and Beau

On a Starry Night

Fun Things to Make and Do From Dusk Until Dawn

Kate Hodges

WHITE LION PUBLISHING

Contents

Introduction

Night time is the right time for... having fun! Coming inside when the sun goes down or firing up the tablets as soon as the curtains are drawn always seems a tremendous waste of precious mucking-about time. Playing out after the sun is down (or before it has risen) is exciting; it focuses senses and sharpens sensations. In the dark you'll hear more, feel more, even, um, smell more. Games are more thrilling, crafts take on different dimensions, and creative play becomes more alive.

Dare to throw off restrictions for night-time play. You'll find ideas here to suit wintry evenings indoors, balmy summer nights camping, fireside parties on the beach, or dark autumnal afternoons. And the great news is that all of them can be adapted to your location and the season – who says you can't sit round a bonfire in winter or make a horror house in spring? I've included things do that will work whether you live in a city, a town, or in the country. Some will be all-night odysseys, others can be done and dusted in the gap after tea-time and before bed.

All of the activities have been tested by real-life children – my twins Arthur and Dusty and their friends, as well as a gang of kids from the adventure playground where Jeff, who took the stunning photographs, works, plus the sparky kids at Greenwich's Christchurch School, in their beautiful community garden. Some ideas will suit older children, some are perfect for toddlers and many are scalable according to age; we tried to include activities that work for families of all shapes and sizes, and of all physical capabilities. This book is designed to be read and used by everyone, from grandparents to the smallest children – pick it up, have a flick, and you're sure to come across something that appeals.

Of course, there will always be a time for sitting cosied up in front of the TV, but I do hope this book will inspire you to be creative with your evenings as a family.

After-Dark Essentials

Some activities require nothing more than a warm jumper and enthusiasm, but others need more preparation. During the making of the book, I found it really useful to pack a rucksack full of essentials that could be used across activities, as well as a stock of extra fun bits and bobs. This meant we could get ready quickly and easily whenever the mood took us and head straight outside for fun. Here's what we packed:

Really Useful

- A good-quality focusable torch
- Red cellophane *(for covering the torch to preserve night vision)*
- UV (blacklight) torch *(find one online)*
- Head torch
- Matches or fire-lighter
- Battery-powered fairy lights
- Warm jumpers
- Sturdy boots or shoes
- Fluorescent paints or powders *(there's a recipe for home-made glowing paint on page 71)*
- Small, stand-up work light
- Small plastic groundsheet or old plastic bags *(for sitting on)*
- Rug or blanket
- Water bottle
- Sweets

Not Essential But Fun

- Binoculars
- Fluorescent make-up
- Laser pointer
- Glow vitamins (see page 37)
- Fog machine
- Electroluminescent wire and battery pack

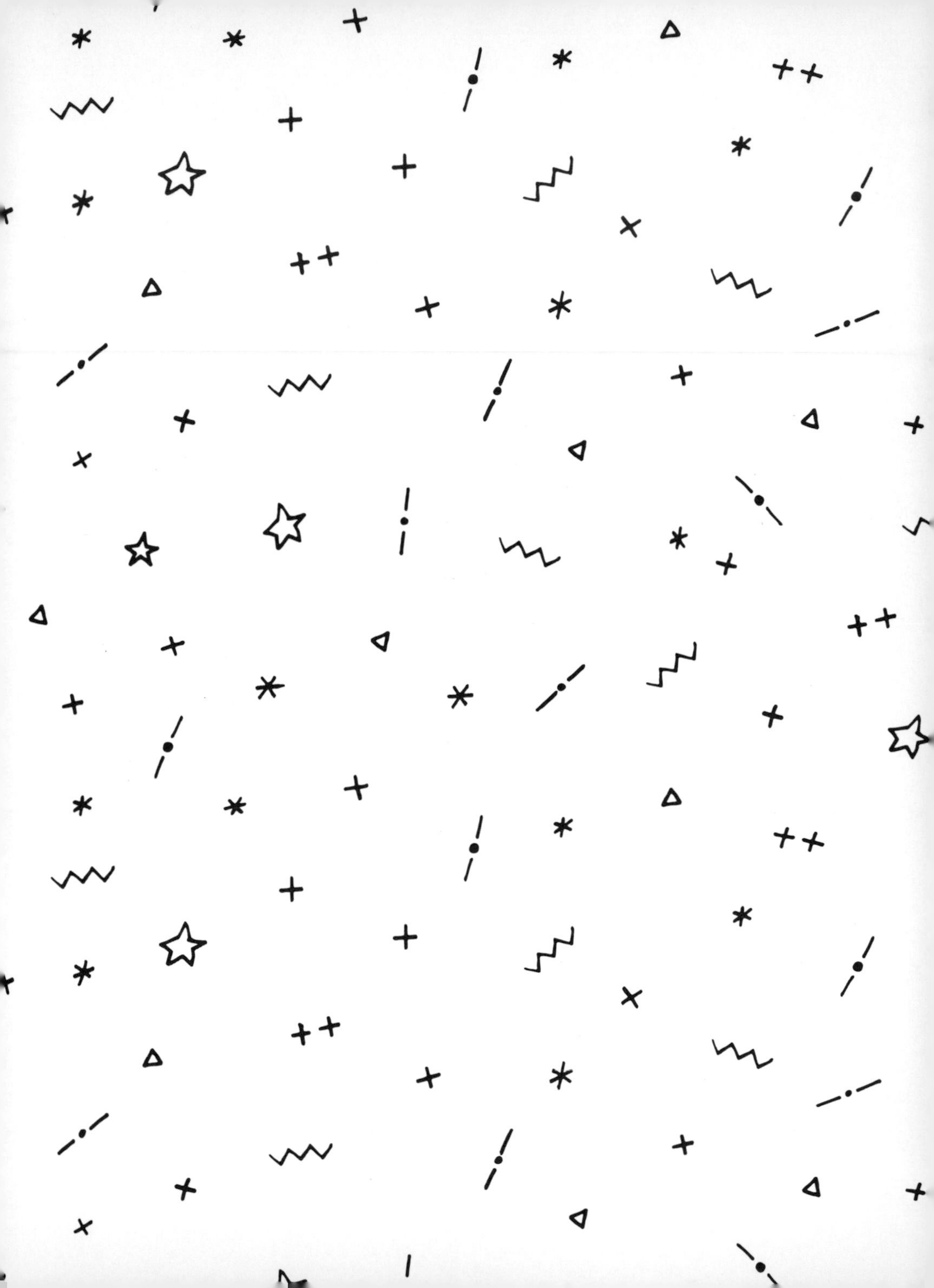

Outside After Dark

Back Garden Cinema

Watching a film outside brings an extra dimension to movie-going. It's sociable, fun, and, with a little preparation, incredibly cosy

Autumn is a great time for movie nights – the evenings get darker a little earlier than in high summer, but it's still warm enough to sit outdoors.

For a scaled-down option, gather round a laptop or tablet.

The Projector

- A really good, bright projector (we recommend something over 3000 lumens) can be expensive. But think creatively: perhaps a few families could pool together to buy one, you could find one second-hand, or you may be able to borrow one from your nearest Library of Things or community group.
- You'll need to hook up your projector to a DVD player, TV or laptop in order to stream your movie or watch your DVD.
- A speaker — wireless or hard wired — will ensure your sound is crisp and loud enough to hear. Most importantly, practise setting everything up beforehand.

The Screen

- You can buy specially made screens, and they do a tip-top job, but there are plenty of cheaper alternatives. Paint a wall white or hang a sheet from a washing line.

The Seats

- You'll need something soft and waterproof. Basic seating could be outdoor chairs, a bench made from bricks and a plank, camping chairs, or a groundsheet with cushions scattered on it. Or, for the ultimate in luxury, sun loungers.
- Why not inflate beach toys and rings for super-cool, kitsch seating?
- Create different heights and options for kids and adults. Bring out your softest blankets, pillows and duvets for maximum cosiness.

The Decor

- Drag a low coffee table outside to put your drinks on, or make your own from pallets.
- String fairy lights around the garden (turn them off when the film is on) or scatter home-made lanterns (see pages 114 and 117) or tea lights. Set up a tent to lounge in.
- Light insect-repelling candles if there are midges about - nothing spoils a film like bites (unless you are staging an interactive screening of *Jaws*!).

The Refreshments

- Popcorn is essential, and it's cheap - and fun - to buy popping corn and make it on your stovetop in a pan. Don't forget to add butter, sugar and a pinch of salt, or try adding herbs, lemon or cocoa powder. Put it into bowls, or package it up individually in paper bags (decorate them yourself).
- If you're using paper cups for drinks, why not customise them too? You could come up with a catchy name for your cinema and design a logo for it. If it's a little colder, keep things toasty with hot chocolate or instant soup.

Why not try a wintry film screening?

Keep the film short, wrap everyone up in their cosiest gear or sleeping bags, and watch something appropriate – *Frozen*, anyone?

Customise Your Bike

Going for a moonlit cycling adventure is even more fun if you zing-up your ride

A simple string of battery-powered fairy lights strung around your handlebars and frame transforms your two-wheeler into whatever you want it to be, from a space-diving rocket to a fairy chariot.

YOU WILL NEED

A selection of the following:

- Battery-powered fairy lights
- Fluorescent stickers or tape
- Light-up valve caps
- Spoke lights
- Wheel lights
- Electroluminescent wire
- Fluorescent bar tape
- Fluorescent paints – paint pens work brilliantly
- Old torches
- Old toys
- Old battery-powered radios
- String, rubber bands or cable ties

1. Wrap UV tape around your handlebars – make criss-cross patterns for a fun effect.
2. Cover your bike in stickers or paint it in fluorescent colours. We love random patterns, but stripes also work well.
3. Use lights to bring your bike to life. Customise your wheels with spoke lights and light-emitting valve caps, or line frames and wheel rims with electroluminescent wire for ultimate flash. For a softer look string fairy lights around your bike, securing them with string or rubber bands.
4. Recycle! Go wild with old toys and torches. Make your bike super-wacky by strapping ancient dolls to the handlebars or dinosaurs to the back of your seat. For tunes on the go, use cable ties to secure a radio to your rear rack, and turn it up loud!

Check with your parents or carers before customising your bike!
Stay Safe!
Keep any trailing wires secured well out of the way of spokes, gears and brakes.
If you intend to ride on a road, never obscure your front or back bike lights.
Make sure everything is secured tightly and never fiddle with decorations as you ride.

Moongazing

Take in the greatest spectacle in the night sky

Our very own satellite puts on impressive shows all on its own, and even the youngest sky-watcher can spot it floating serenely in the sky.

You don't need to make a special trip, simply watch the moon wax and wane (get bigger and smaller) on winter walks home from school, or out of a window before bedtime. See it shining in the afternoon sky or rising over the trees at dusk. Spot it when it's a fat, buttery circle or a shard-thin silver crescent. Watching it grow and reduce over the month connects you to the lunar cycle, to the seasons, to our place in space.

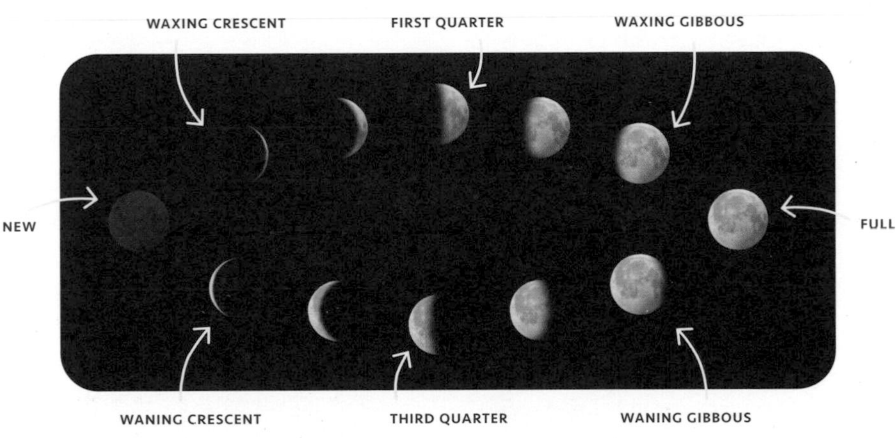

Some moons are extra special. See which ones you can spot:

Blood Moon

Glowing red, you'll see this dramatic spectacle during a total lunar eclipse (when the Earth travels between the moon and the sun, throwing the moon into shadow).

Harvest Moon

You may spot this luscious, full, bright moon in the early autumn. Before electricity was used for lighting, farmers relied on this moon's light to harvest crops as the days grew shorter.

Supermoon

When the moon swings closer to the Earth it looks a little larger. Astronomers call this a perigean full moon.

Blue Moon

Sad news, the Moon doesn't turn navy. This is an 'extra' moon – a second full moon in a calendar month. This happens roughly every two-and-a-half years, and inspired the saying, 'once in a blue moon'.

How to Build a Fire

Making a fire that flickers into life is a satisfying experience, and an indispensable life-skill

YOU WILL NEED

- Tinder – *this is something that will burst into flames easily and needs to be dry and light. Leaves, hay, bark shavings, tiny twigs, lint, newspaper or flammable cloth will all work well.*
- Kindling – *bigger than tinder, but smaller than logs, this amplifies your fire from a whimper to a mighty roar. Sticks the size of pencils are perfect.*
- Sticks of various sizes
- Logs
- Matches, fire lighter, or a fire steel
- Large stones, bricks, or a spade

1. Find a safe place to build your fire, away from tents, walls, overhanging trees or paths.
2. Build a safe hearth. Use large stones or bricks to mark out a clear area, or dig a shallow pit with a spade. Alternatively, clear an area of bare earth, getting rid of twigs, leaves or dry grass.
3. Place tinder in the middle of your hearth.
4. Use kindling to form a teepee shape around the tinder, leaving a 'door' or small opening.
5. Place some larger sticks around the kindling to create a triple-layered structure.
6. If it's windy, get everyone to huddle round, or set up a barrier to keep the breeze from blowing out the baby flames.
7. Poke a match or lighter through the 'door', and ignite the tinder. This should spring into flames, light the kindling, and then set fire to the bigger sticks. Wait until it's burning strongly then add thicker sticks and finally logs.
8. Eke your wood out; if you want to keep your fire going for a few hours, or cook on it, you'll need to keep adding fuel, so don't pile it all on at once.

ADULT SUPERVISION NEEDED!

- **Never leave children unsupervised near an open fire**
- **Ensure all flammable items are well out of the way**
- **Long hair should be tied back and loose clothing removed**
- **Always have a bucket of water on hand to extinguish stray flames**
- **Use earth or water to make sure that your fire is put out properly at the end of your session**

Cooking on Your Fire

It's most efficient to cook in glowing ashes, so let your fire burn for at least thirty minutes before you start. You can use pans, tripods and trivets, but the easiest way to bake is barbecue-style, on a grill placed on rocks, or, most simply, in the fire ashes or on sticks held over those ashes.

Stick Food

A basic way to cook, but lots of fun. Use skewers or sharpened sticks to toast and heat food.

YOU WILL NEED

- Wooden skewers or sticks
- Snacks

1. Soak the stick or a wooden skewer in water for an hour or two to reduce the risk of it burning.
2. Poke your stick or skewer into the chosen food and make sure it's secure. Marshmallows are the classic stick treat, but hot dogs and cheese sandwiches also work well. You might even try baking an egg using a scraped-out half-orange strung on your stick, basket style. Hold your impaled snack just above the glowing ashes where the heat is even and strong.
3. Try toasting apple slices, pineapple rings or peaches. Brush the fruit with honey or butter for extra caramelisation.

Baking in the Ashes

Use the ashes of the fire to cook your dinner while you tell fire-side stories. Always take care to remove your food using barbecue tongs or a long-handled fork and set aside until cool enough to handle.

YOU WILL NEED

- Heavy-duty aluminium foil
- Tongs or a long-handled fork

Baked Potatoes

A dream to cook in camp fires.

1. Cut them in half and fill them with butter, then wrap tightly in heavy-duty tin foil and place in the fire.
2. Use tongs or a rake to pull some embers over the top of the potatoes so they cook evenly and allow to cook for 45 minutes for white potatoes, 25 for sweet potatoes.

Sweetcorn

If the cobs are still in the husks:

1. Leave the husks on and soak the corn in water for an hour before cooking.
2. Place in the ashes of your fire for about 20 minutes.

If the cobs are not in husks:

1. wrap in foil and cook in the ashes for around 10 minutes. Slather your cob in butter and enjoy.

Roasted Vegetables

Sliced mushrooms, tomatoes and peppers all work well.

1. Make a tin foil pouch by pinching a piece of foil together.
2. Place the veg into the pouch, then drizzle with olive oil.
3. Place the pouch into the embers and cook for 6–7 minutes.

Fish

Freshly caught mackerel is particularly delicious cooked this way, but any firm or oily fish works well.

1. Place the fish onto a piece of tin foil and season with pepper, salt, lemon, herbs and butter.
2. Pinch the foil around the fish to seal and cook in the ashes for 6-7 minutes.

Chocolate Bananas

Sweet and gooey goodness.

1. Keeping the skin on, cut slits in a banana.
2. Push chunks of chocolate into the slits then wrap in tin foil and cook in the fire's ashes for 7 minutes.
3. Serve with cream and mini marshmallows or peanut butter.

Campfire Damper Sticks

Sweet bready treats are easy and quick to make over a firepit

As this recipe is so simple, you can take the ingredients out and make it on the hoof. Alternatively, mix the ingredients in your kitchen ahead of time, and take the dough out in a container. Try them camping, on the beach, in the garden, even round a winter fireplace.

YOU WILL NEED

- 500g (1lb 2oz) self-raising flour
- 80g (3¼oz) caster sugar
- 200ml (7fl oz) water
- 8 skewers or sharpened sticks
- Tin foil
- Mixing bowl
- Wooden spoon
- Jam, chocolate spread, butter, to serve

1. Wrap the skewers or sticks in foil.
2. Put the flour and sugar in the bowl and mix.
3. Slowly mix in the water until the mixture forms a soft dough. Set aside for ten minutes.
4. Divide the dough into eight pieces, then form each piece into a sausage shape.
5. Wind the dough sausage shapes around the sticks and press to make them secure.
6. Hold the sticks or skewers over a fire, turning now and again until each side is golden brown (about 10 minutes).
7. Serve with jam, chocolate spread or butter.

ADULT
SUPERVISION
REQUIRED
!

Watch the Sun Rise

It's an everyday miracle

Witnessing the glow of the sky as the dawn breaks can be a surprisingly magical and emotional experience, particularly if you do it with your family. At this time of day there's less pollution in the air, so the sun has sharper edges and the colours are brighter and more focused.

Check the weather forecast and try to select a morning when a ridge of high pressure hangs over the country and there are light winds, as both of these elements will increase the chance of a spectacular-coloured sky.

Choose your spot from which to watch wisely – you will need to be facing east, and higher places will ensure a better view. Looking out over open seas mean that the light and colours are reflected, creating an even more breathtaking vista.

Check sunrise times and be sure to add in a few extra minutes if you're up high – the sun will be visible sooner. The colder months are the ideal sun watching time, as the longer nights means it's less of a struggle to get out of bed on time.

Add meaning to watching the sun rise by creating a family tradition. Perhaps you could do it on the same day each year; January 1st, on someone's birthday, or even the first morning of the winter holidays.

You might make private wishes or resolutions together as the rays warm the earth, or just let yourselves be transfixed by the colours. And, if the clouds obscure your view, or it's too cold to enjoy, you can come back the next day to try all over again.

Go for a Night Hike

Even familiar trails take on a different character after the sun goes down

Walking at night is a totally different experience to wandering in the day. Your vision is reduced, which means your remaining senses come alive; you might hear rustling, strange animal calls or night bird song; you'll smell more acutely, the scents of night-blooming flowers or damp undergrowth; and you'll have to rely on touch a lot more to find your way, feeling the path underfoot and reaching out for stiles.

Walking a path you know well is a good way to begin your night rambling adventures.

Plan your route in advance so you don't need to fumble with maps or apps as you go and to check for any potential hurdles such as stiles or muddy patches. Your ramble doesn't have to be long as even short night walks are a thrill for small children. You could choose to wander through a nearby wood, a park or scale a hill for night-time vistas and stargazing.

Think about timings; will you set off before the sun sinks, and watch the sunset? Or will you get up early and watch it rise? You might even wake up in the middle of the night for a pitch-black adventure.

YOU WILL NEED

- Torch or headlamp
- Map or phone app
- Snacks
- Drinks
- Walking boots or trainers
- Pocket-sized first aid kid
- Insect repellent
  (depending on location and time of year)

BE PREPARED

- Taking your first steps is exciting. This is it! You're off! Take time to revel in the quiet and listen hard for animals' crunches and rustles. Young (and not so young) children might feel more secure if you hold hands; this is where head-torches come into their own. Be prepared to walk more slowly than usual, which should be no hardship as you'll become even more attuned to the natural world.
- Make regular stops en route to sit, turn off your light sources and revel in the darkness. You might look for constellations or take in a view of your town's streetlights, pointing out familiar roads or schools. Perhaps you'll discover the joys of companionable silence and find some time to be alone in your thoughts.
- The last stretch might be tricky, but sing a song or play a game to keep spirits high. A well-timed snack is everything, and the encouraging rustle of a sweet packet can help give those with smaller legs motivation to reach the top of that hill.
- Plan your arrival treats in advance, perhaps you'll have hot chocolate or cinnamon-spiked warm milk to welcome you home and celebrate your achievement. What an adventure.

It's a beautiful time to walk – you might spot wildlife, meteors or just enjoy the full moon.

If daytimes are really hot, it's even a practical way to enjoy the outdoors with your family; no SPF required!

HIKING SAFETY CHECKLIST!

- **Let someone know where you're going and when you intend to be back.**
- **Check the weather forecast before you go, and don't be afraid to call your walk off if it looks dicey.**
- **Set ground rules at the start so everyone knows what to expect.**
- **Assign each person a number and do regular count-down shout-outs to check everyone's okay.**
- **If someone needs a break, everyone should stop until they're good to carry on.**
- **Make sure you have a fully charged mobile phone with a map app.**
- **Everyone will need a full water bottle, snacks, a torch or headlamp and appropriate footwear.**
- **Make sure everyone is wearing appropriate clothing in layers and has a waterproof jacket.**
- **If there are dangerous creatures in your country, be aware of where they might be lurking.**

Tell a Gripping Fireside Story

With a little practice, you can become a master myth-weaver

☾

Spinning a tale in the glow of a fading fire is a magical way to bond with your family as well as exercising your creative muscles. But the thought of telling stories from scratch can be intimidating. However, do a little preparation, and you'll be pulling sagas from thin air, much to the delight of the rest of your clan.

- Remind yourself that the experience of storytelling is about sharing rather than performing. It should be an enjoyable experience for both you and your listeners.
- Cast yourself as a custodian, keeping the special stories that you find alive through telling them to others. This can help remind you that you are one of millions of people around the world who are sharing stories with each other.
- When trying to make up your own stories, start with simple ones. Adults, you will be surprised how interested children are in memories from your childhood. Kids, you can tell tales of what happened at the park, or stories based on games you've made up at school.
- Even the most unlikely of topics can spark a wonderful tale; we know of one dad who based every one of his stories on video games he played as a kid; the giant monkey who stole a princess, the ball who ate dots and was chased by ghosts, and the saga of Link and his adventures in Hyrule.
- Ghost stories are a great place to start; as long as your initial situation is relatively normal, but there's an unexpected shiver in the tale's tail, you're guaranteed to entrance your listeners.
- Use a set of story cards or dice to give prompts for your tales; you'll find that everyone will want to join in, which creates a warmly communal experience.

Yarn-Spiration!

Fairytales

You can't go wrong with a princess, a prince and a witch. But try turning traditional storylines on their head to create something new; an evil princess, a cowardly prince, a funny wizard, or a dragon who can't breathe fire.

Mythical Creatures and Ghosts

Be inspired by traditional tales of mysterious creatures; selkies, the seal people who transform from the doe-eyed mammals into humans and live for years on land; mermaids; vengeful spirits; werewolves; wild dogs; witches and vampires. Search for folk stories set near your home or where you are on holiday and tell your own version for added local spice and scariness.

Animals

Make like Greek fable writer Aesop, American folklore genius and Brer Rabbit creator Uncle Remus, or the West African tellers of the Anansi stories (tales featuring a talking spider), and base your stories on a group of talking creatures.

Inside After Dark

Create a House of Horrors

Spook your friends at a ghoulish party

Every Halloween, our home transforms. Admittedly, the place is pretty spooky, even in the day, but in late October, it becomes positively terrifying. With a little preparation, and a generous helping of wicked fun, even the tiniest of flats can be transformed into a ghostly experience, ready to receive victims, sorry, guests.

This is an activity that you can scale up or down, depending on your ages, temperaments and capacity for terror – keep things brightly-lit and cartoon-like for smaller children, mysterious and chilling for those a little older, and positively hair-raising for teenagers. It doesn't have to involve a lot of expense and running out to buy ready-made decorations – we've found that home-made props are often a lot more sinister!

Make a Plan

Think about how your haunted house is going to work. Planning ahead will reap spooky rewards.

- How old are your guests?
- Will it be a walk-through, or are you going to hold a party?
- Will you have a theme?
- What props and items do you already have handy that you can re-use and repurpose?
- Which areas of the house will you transform?

Outside Decorations

If you can intimidate your guests before they even knock, you're halfway towards giving them the scare of their life. Lighting is crucial for every area of your bone-rattling house, but it's particularly effective if used outside. If you have a garden, use it! Hide lights (outdoor fairy lights or lamps) behind branches to throw weird shadows onto pathways and walls. Use a smoke machine to create a foggy, weird atmosphere. Place flickering LED lights or throwies (see page 130) in dark places to resemble eyes. Make cardboard gravestones with funny inscriptions.

Use those trees and bushes! Create your own spooky figures from old mannequins, decrepit dolls, old plastic bags, and worn-out clothes or scraps of material and hang them from branches, or prop them up against fences or walls. One of our most successful spooks was a simple-yet-sinister scarecrow, with a sackcloth head, tatty old hat and sewn-up mouth, slung on a cross pole. Light them from below to throw more shadows and make them even more scary.

Think about your windows – cut out silhouettes of ghosts or skeletons and stick them to the inside of the glass, and make tattered curtains from sackcloth or shredded cheesecloth, stained in a pan of black tea. Illuminate the windows from inside with flickering LED candles or torches.

Inside decorations

Once your visitors are inside, shut the door with a firm click. They're YOURS! Think about setting all your guests' senses tingling.

Sight

Lighting is key. Keep the house dim – use lamps, flickering LED tea lights, coloured torches and lace dresses or silky scarves to cast shadows and create atmosphere. Use wool and rip up old sheets to create your own eerie silhouettes. If you have shop-bought decorations, highlight them from below. Hang old toys from the ceiling – even the most cuddly of teddy bears can look terrifying in the dark! Want to keep certain areas out of bounds? Block them off with giant spiderwebs made from wool.

Hearing

Record a spooky soundtrack; screams, bangs, hysteric laughter and creepy songs on a phone, or search online streaming services for 'horror sounds' and play them as loudly as you can on a stereo.

Touch

Create strange textures from the ground up by carpeting the floor with foam or sponge and drape satin sheets or old blankets on the walls. Let wet strips of material dangle down from door frames and brush against your friends' faces.

Smell

Scent can help create a scary atmosphere. Spray puffs from ancient bottles of old-lady perfumes to conjure up ghostly presences or grab some mouldy bits from the compost heap and put them in a bucket. Bleuuurgh.

Characters

Using real people is perhaps the most effective scare tactic. Rope in friends and relatives to play parts – for younger guests, this might involve an adult dressing up as a friendly fairy 'guide' or harmless witch, but for older kids and teenagers, this is where you can ratchet things up. Costumes don't have to be explicitly horror-based – ghoulish grannies, zombie mechanics, phantom brides and generic shrouded figures are all equally terrifying when they're lurching out at you from behind a closed door!

You don't have to stick to horror housing in October; spooky birthday parties are a lot of fun at any time of year.

Mad Scientist Specimen Jars

This display makes old plastic toys look terrifying and is so easy to make. Figures that work particularly well include rubber animals such as bats or snakes, monsters, dolls heads, and Halloween-specific goodies such as severed limbs or eyeballs.

1. Put your chosen toy inside a jar.
2. Fill the jar with water and add a couple of drops of food colouring – green or yellow are particularly effective. Stir and ensure the figure is at a good angle. Close the lid.
3. Light the jars from behind using LED tea lights or a lamp.

YOU WILL NEED

- Old jam jars with lids
- Old plastic and rubber toys
- Water
- Food colouring
- LED tea lights or small lamp

Be Safe!

- **Set ground rules for your characters – no touching!**
- **Create a safe, well-lit retreat zone for anyone who finds things too scary.**
- **Make sure flooring is secure and hanging strips of material will come loose if tugged.**
- **Never use real candles or fire.**

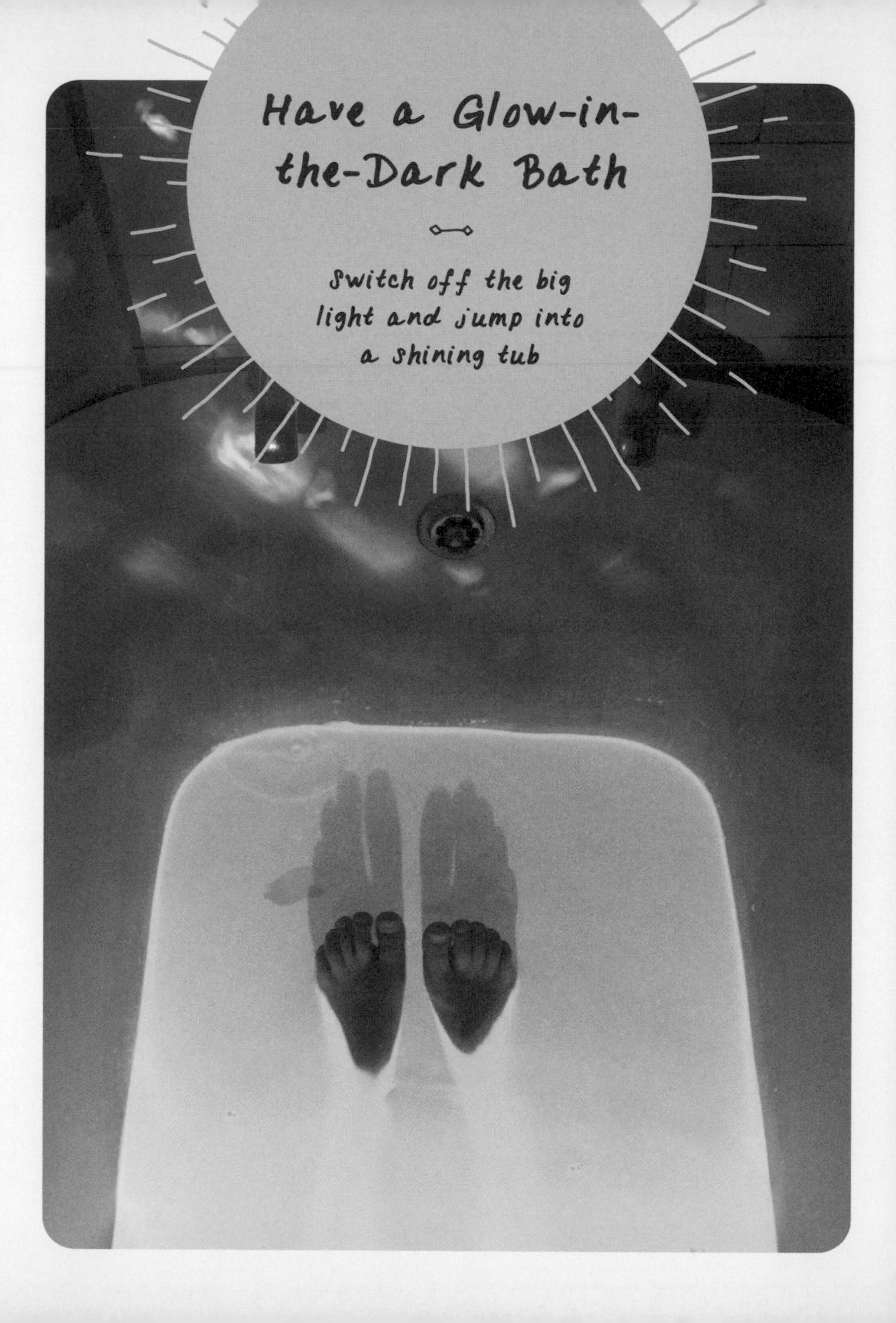

Have a Glow-in-the-Dark Bath

Switch off the big light and jump into a shining tub

Adding a sensory twist to bath times turns your tub from dull to disco. Sit battery-powered tealights around the edge of your bath, or set up a mirror ball in your bathroom for extra illumination. Turn out your overhead light and revel in the soothing, warm glow.

There are a few different ways to create a party bath. The easiest method is to add a couple of glow-in-the-dark necklaces or glow sticks to the water, but in an age where we're all trying to cut waste, it's difficult to justify using anything non-recyclable. To create your own fluorescent, safe, bath water, you'll need glow vitamins. Use B-complex or B-50 vitamins with the following combination of ingredients for maximum shine; thiamin 50mg, riboflavin 50g, niacin 50mg, vitamin B6 50mg, folic acid 400mcg, vitamin B12 50mcg, biotin 50mcg and pantothenic acid 50mg. Search online for a brand that suits you.

YOU WILL NEED

- A battery-operated blacklight *(under no circumstances use a mains operated light near water)*
- Glow vitamins (see above)
- Pestle and mortar OR sandwich bag and rolling pin

1. Use the pestle and mortar to crush eight vitamin tablets (or put eight into a sandwich bag and crush them with a rolling pin).
2. Add the crushed vitamins to a glass of warm water and stir to dissolve.
3. Run a bath and pour in the vitamin mixture.
4. Turn off the bathroom lights, fire up the blacklight, and watch the bath start to fluoresce.
5. Get your kids in and play! The water is safe to splash in but you may need to rinse off the orange colour afterwards.

Hold the Best Rave

Turn your living room into a disco

Dancing makes everything better. And while it's fun sticking some tunes on and having a leap around, it's even more exciting making things into a proper rave, and raves are most fun when it's dark. You'll need space for this, which isn't always easy to find.

Sounds

Whether you're streaming a playlist, rummaging through your parents' CDs, or going old school and playing records or tapes, your choice of songs is important. Tailor your playlist to your guests – small kids love TV themes they know and souped-up versions of nursery rhymes, whereas older kids will be a little more street smart. Making a streaming playlist in advance keeps things simple, and you have a world of music to choose from. Don't forget to throw in a few curveballs to get your dancers smiling; old songs familiar from film soundtracks or crazy cover versions of well-known tunes. You might even want to plug in a mic so you can sing along with records, freestyle over the top of them, or direct games.

Setting up your stereo should be relatively easy, but practice before the big day; turn it up as loud as you can without annoying anyone. If you're outside, you'll need to figure out how to power it – extension cables, batteries or even a generator! If you don't have a big stereo, that doesn't matter – a phone with an external speaker or even a radio will do the job just as well.

You'll need space for this, which isn't always easy to find. If your front room isn't big enough, why not try borrowing a bigger space (community and arts centres often have rooms that can be hired), or going mobile and moving things outside to a garden, park or beach?

Lights

Lights are incredibly important: they create atmosphere and make things special. We found all of ours at boot fairs or charity shops – keep an eye out for bargains – but do get them checked by an expert before you use them. To add even more spice, we add haze using our smoke machine (also found at a boot fair), which softens harsh lights and builds excitement. If you don't have proper disco lights or a smoke machine, don't worry. Fairy lights look great, especially those that flash in different patterns. Or improvise with torches. Coloured or LED colour-changing bulbs make great rave lights. Dig out old light-up toys and hand them out to your friends to wave around.

Extras

Make fluorescent decorations using paints, tape or UV stickers, and rig up your blacklight to illuminate them. Try creating large-scale murals on cardboard boxes – flowers and slogans work well. Wear your UV glow clothes (see page 110) and chuck a few throwies (see page 130) around for maximum, eye-popping effect!

WARN YOUR NEIGHBOURS!

- **If you are going to play your music loud, it's only polite to let the people on your street know. Invite them along, or at least tell them when the party is going to be over.**
- **If you're in a public spot, be aware of how much noise you're making and turn it down or move along if asked. If you're planning an outdoor party, check the weather forecast – if there's any chance of rain, call it off – electrics and downpours don't mix well.**

Shoot a Spooky Film

Tell your own terrifying tale

This is an activity that's scalable to any age or family shape. Younger children will love running around with sheets over their heads, then watching themselves on video. Older kids can develop storyboards and scripts, and create their own costumes, while teenagers will revel in making something truly horrifying to give real chills. Most mobile phones are able to shoot and edit video, so making your own scary film is just a case of having one to hand.

YOU WILL NEED

- Mobile phone or digital camera with video shooting capabilities and storage
- Costumes
- Props
- Torches or lights
- A vivid imagination

1. **Have a rummage through your dressing up box or old clothes pile**, and base your characters and storyline on what you find. Use shop-bought costumes or find hats, scarves, long coats, sheets, masks and make-up to create terrifying monsters, ghosts and ghouls.
2. **Find props.** You might already have a stash of fancy-dress accessories, but try repurposing toy spiders, old dolls, tools, kitchen implements or game pieces. Remember, anything can be haunted or possessed!
3. **Create your story.** You might like to come up with a rough plot then improvise, or work on a more solid script. Discover how you work best!
4. **Draft in friends and relatives** to act in your film. It's good to have a wide range of ages; use baby sisters or brothers, grandmas or grandads.
5. **Your story may depend on whether you have access to editing software.** You can find powerful apps that enable you to cut and paste video on a phone to make a film, or upload your footage to a computer and use a basic program such as iMovie. Alternatively, shoot a short film all in one take.
6. **Think about lighting.** Use darkness to create atmosphere and lights to cast shadows. Test your set-up – is it light enough to see what's going on? A little confusion is fine – in fact, it adds to the eerie feel.
7. **Shoot your film.** It's an old cliché, but scaring your audience is as much about what you don't show as what appears in the final cut. To really weird people out, try to avoid using too much footage of your villains, ghosts, or monsters and instead hint at their terrifying nature. Mix close-ups of your actors with longer shots and don't be afraid to leave plenty of time where no-one speaks. Have an unexpected ending.
8. **Add in music and sound effects.** Search online for libraries or make your own using creaky instruments and things that go bump, boing and splat. Try rattling rice on a tray, squishing ripe fruit, hitting rolled-up newspapers with soft wooden sticks, rubbing plastic surfaces, dragging plastic tables or snapping Styrofoam or pasta to make creepy noises.
9. **Add in the title** and credits using your editing software. Your audience will want to see who made and starred in the film!
10. **Hold a screening**; perhaps in your Back Garden Cinema (see page 12)

Ultraviolet Body Paint

Bring an extra glowing dimension to a party, or simply liven up a boring evening with a creative, skin-art session

Fluorescent body paint glows brightly under your blacklight. A blacklight bulb or stand-alone lamp works best as it lights up the whole room, but using a torch to shine on to your face or body parts can be equally effective.

YOU WILL NEED

- Body paint in stick or liquid form
- Blacklight torch or bulbs

OPTIONAL

- Make-up brushes
- Fluorescent hair gel
- Fluorescent mascara
- Wigs
- Hats & scarves
- Fluorescent glow lipstick

1. **Choose how much of your body you want to paint.** Wearing a swimming costume will give you the largest expanse of skin to work with, but you can create dramatic effects using just your face, or hands, or even your feet. Work under the UV light so you can see exactly what you're doing – the paint doesn't show up well in natural or normal bulb light.
2. **Now choose your design.** You can stick to an overall theme or get creative in abstract styles. Aim to leave some dark areas for contrast – sometimes restraint can be very effective. Your eyes and lips can look a bit strange if they remain unpainted – find specialist mascaras and lipsticks online to outline them.

Cheat!
If you haven't got UV bodypaint, try experimenting with highlighter pens on your skin! Keep them well away from children's mouths.

Ideas To Try

- **Skeletons**: mark out ribs, a skull shape, hand bones and even a heart. Be inspired by the colourful dots and petal-shapes of traditional Mexican Day of the Dead looks. Add a top hat or scarf to finish things off.
- **Spirals**: paint swirls and whirls on your cheekbones, your shoulder blades and your thighs. Sculpt your hair into wild shapes using UV hair gel.
- **Polka dots**: easy peasy to do with fingers, and this design works equally well on your body or face. Wear a crazy wig that glows for clown-style fun.
- **Animal**: paint your body tiger or cheetah style, and your face with a nose and whiskers to match. Make a tail from material that glows under UV lights and spin it around.
- **Stripes**: mix and match colours in simple stripes down your face and across your arms. Go all-over or stick to a few lines on your cheeks and nose.
- **Abstract**: let your imaginations run amok as you create splodges, splashes and whirls. These designs look wild under UV light. Squidge the paint up into your hair, mix colours and randomly swirl them around for mind-blowing effects.
- **Got an awesome pattern on?** Great! Try creating some crazy dances – your moves look even better highlighted in UV. Co-ordinate skeleton jigs or go freestyle – watching the neon colours bob and flow is mesmerising. If you want to preserve your look, take selfies or film yourselves. Upload them for the ultimate, attention-grabbing posts on social media. Rub and swirl the paint around for one final, strange, abstract look. Then jump in the shower!

Create a Magic Ritual

Mark the seasons with ancient traditions

Taking part in ceremonies with your family, whether they are magical, religious, traditional, or conceived on the spot by active imaginations is a great way to bond, get creative, be mindful and even tackle bigger, emotional subjects as a team. Enacting them in the dusk or dark makes them even more dramatic and special.

Spring Dew

The glinting beads of moisture that gather in the grass on the first day of spring are said to have special properties. It's said that the drops will protect anyone who washes their face with the water from pimples and wrinkles for the next year, and that the magical dew will bestow the gift of beauty. Even Samuel Pepys, writing in 1667, mentions the tradition. Set your alarm clock for before sunrise and head out in the dark to rinse your face and welcome the Beltane (the Celtic spring-time festival) sun.

Summer Moon Bathing

An Indian system of medicine, Ayurveda uses the balm of the moon to calm Pitta Dosha (anger, or too much heat, in the body). So why not see if lunar rays chill your family out? Sit outside in a circle during a full moon, holding hands, and breathing deeply. Burn incense or a scented candle for extra atmosphere and relaxation. Lose yourselves in the quiet, and reflect on your futures, or quietly set gentle goals for the coming month. Some people believe that objects 'charged' under the moon have special potency – try leaving a glass of water outside overnight and drinking it in the morning to test its superpowers.

Autumn Peel and Peer

The Roman goddess Pomona was known as the apple queen, and her festival fell around November 1st. This might be where the Halloween traditions of bobbing for apples or apples on strings come from. Apple peels and pips have also been used to divine romantic futures; it was said that you could learn the initial of your future husband or wife at midnight on Halloween. To find your future partner's name you'll need a mirror, an apple, a vegetable peeler and a candle. As midnight strikes, peel the apple in front of the mirror in one continuous strip, then throw the strip over your left shoulder. The letter it resembles will be the initial of your true love.

Winter Festival of Light

Imbolc is the Celtic festival that marks the end of winter and the return of the sun. In the Christian calendar, it's known as Candlemas. Marking it with your family is a neat way to observe the turning of the seasons; this is the starting pistol for green shoots, lambing and warmer days. Traditional celebrations involve the lighting of candles and bonfires, so why not hold an outdoor winter party, with fires, singing and dancing? Or follow the Celtic tradition and make a St Brigid Cross – or sun wheel – from sticks, straw, rushes or paper drinking straws.

Get Sensory

UV slime, glowing bubbles, and shining rice

Stimulate senses and explore simple scientific principles with these squishy, floaty, satisfyingly icky projects. Perfect for really young children, but, let's face it, you're never too old for slime and bubbles!

Glowing Slime

No-one can resist the squidgy, icky feel of slime, and glow-in-the-dark goo has extra woo-hoo factor. Even the tiniest child can join in and help make this easy, squeezy version that shines after exposure to light.

YOU WILL NEED

- 150ml glow-in-the-dark glue *(a greenish colour works best)*
- 2 tbsp water
- ¼ tbsp baking powder
- 1 tbsp contact lens solution containing sodium borate and boric acid
- Large bowl
- Light source

OPTIONAL

- Zip-lock bag or container with lid for storage

1. Squeeze the glue into the bowl.
2. Pour the water into the empty glue bottle, give it a shake then add this to the bowl (if you like your slime reeeeeally stretchy, add a little more water).
3. Add the contact solution slowly, taking care not to add too much or your slime will be too hard.
4. Knead. And keep kneading. Then knead some more. This part takes time but is the most fun.
5. Expose the slime to a light source (the sun or a lamp), then turn out the lights and watch it shine!

ADULT SUPERVISION NEEDED!

If your skin becomes irritated, stop handling the slime and wash your hands immediately.

UV Slime

This is a simpler slime recipe that needs a UV light source to shine.

YOU WILL NEED

- UV light source
- 400g (14oz) cornflour
- Water
- Fluorescent powder or paint
- Large bowl

1. You'll need roughly twice as much cornflour as water but experiment to get the desired consistency.
2. Put the cornflour into the bowl.
3. Add the paint or powder to achieve the desired colour.
4. Turn out the lights and turn on your UV light source.
5. Slowly add water while mixing until the mixture reaches the desired consistency. This is the best part! Will you go for a solid texture or something super-gooey? See what happens when you stir the mixture.

Glow-in-the-Dark Bubbles

Make bubble time even more magical with glowing, shining versions.

YOU WILL NEED

- Jar
- 2 cups washing up liquid
- 1 cup water
- 1 tsp glycerine *(optional)*
- Fluorescent powder or paint
- UV light source

1. Pour the washing up liquid into the jar.
2. Add the water and stir.
3. For a stronger solution and bigger bubbles, add the glycerine.
4. Add fluorescent powder or paint and stir again.
5. Turn on your UV source and blow in the dark!

Shining Rice

Tiny children will love the swooshing sounds and soothing sensations of glowing dry rice.

YOU WILL NEED

- Rice
- Bag
- Fluoroescent or glow-in-the-dark powder or paint
- Bowl
- UV light source *(if using fluoroescent paint)*

OPTIONAL

- Essential oil of your choice

1. Pour the rice into the bag.
2. Add enough glow-in-the-dark or fluorescent paint to coat the rice.
3. Add essential oils if you'd like the rice to smell as well as look good.
4. Shake until all of the rice is covered in the paint or powder.
5. Pour the rice into the bowl and allow to dry, stirring gently every so often so the rice doesn't stick together.
6. Turn on your UV source or expose the rice to light.
7. Turn off the main lights and see the rice glow.

Have a Glow-in-the-Dark Feast

Discover which foods and drinks will glow-up snacktime

Glow On!

Bank notes, sun block and petroleum jelly also shine under UV light.

This fun activity allows you to find out which food and drink glows in the dark – half experiment, half excuse to stuff your face. Use white plates to make the food stand out even more and raid your kitchen to find some cool colour-changing food!

YOU WILL NEED

- UV light source
- Transparent containers
- Plates

Tonic Water

This bitter-sweet fizzy drink might taste a little weird, but it has an ingredient that makes it very special; quinine. This compound comes from the bark of the cinchona tree, is used to fight malaria, and, as it contains phosphors (a chemical substance that fluoresces), glows blue in the dark. Try pouring a straight glass of the liquid and shining your UV light source on it. Cool eh? But you don't have to stick with using it as a stand-alone drink. Try using it in place of water to make glow-in-the-dark jelly or freeze it to make glowing ice-cubes.

Effervescent Vitamins and Energy Drinks

Vitamins give you get up and glow! Vitamin A and the B variations thiamine, niacin and riboflavin all shine bright yellow under a UV light. We've found that a vitamin B tablet high in thiamine (ours is called B-50 complex) is the brightest, but why don't you try experimenting? Grind up tablets, dissolve them in vinegar and see what colours they fluoresce. Some energy drinks have added vitamins that make them shine under UV lights, too, but many have lots of added sugar and are caffeine-heavy, so find brands suitable for under 18s.

Honey

Like these vitamins, honey contains flavins, which makes it glow golden yellow. Try spooning some into a glass pot and holding it up to a UV light source.

Salad

Chlorophyll glows red under UV lights. Make a freaky salad by assembling fresh vegetables that shine brightly. Lettuce, spinach or coriander will glow red, as will peppers. Squash appears yellow. Add slices of pineapple for a vivid blue glow.

Olive Oil

Give an orange kick to your salad with a drizzle of olive oil or put it into a glass and shine UV light through it for maximum effect.

Turmeric

Sprinkle turmeric powder over food for a yellow-green shimmer. Alternatively, try dissolving it slowly in water or vinegar and seeing what colours and eerie effects you can create.

Rock Salt

Some minerals fluoresce, including rock salt.

Bananas

Shine a UV light on a ripening banana and see the outer edges of its black spots shine blue.

Milk, Ice-Cream, Yoghurt

These dairy products have a cheery yellow glow in the dark – more like custard!

Eggs

Hard-boil some eggs for a purpley-red tinged treat. Crack them open, and the white glows yellow.

White Sight

Many foods don't actually glow but instead reflect light. Use pale foods as a base for your glow foods to make your feast even more spectacular: marshmallows, crisps, pasta, rice, white bread.

Rainbow Fire Pinecones

Use pinecones to add colour to your fireplace or camp burn-up

Not only do pinecones burn quickly but steadily (you can use them as kindling) but, if you prepare them beforehand, they will create a colour-filled glow.

YOU WILL NEED

- 10 pinecones
- 1 bucket
- Something heavy-ish (a lid, plate or dish) that fits into the top of the bucket
- 2 litres (3½ pints) hot water
- Newspaper
- Choice of rainbow-maker *(see below)*

1. If the pinecones are damp, leave them out to dry and open up a little.
2. Carefully pour the hot water into the bucket and dissolve your chosen rainbow maker in the water.
3. Put the pinecones into the bucket and place the lid on top to hold them under the water. Leave to soak overnight.
4. Remove the pinecones from the solution and put them on newspaper to dry for at least three days – putting them in an airing cupboard will speed things along.
5. Ask an adult to add your pinecones to a campfire or stove and watch them glow.

Rainbow-Maker

For young children

- 1 mug of salt (for a yellow glow)
- 1 mug of baking powder (for an orange glow)

For older children

- 1 mug of epsom salts (for a white glow)

Stay Safe!

- **Stay well back from the pinecones as they burn, as they can sometimes 'pop'.**

ADULT
SUPERVISION
REQUIRED
!

Write Secret Messages

Keep things hush-hush with invisible inks

Sometimes you don't want everyone to read a note you're sending to a friend. You may have some news you don't want to share, you might be arranging a secret rendezvous, or you might, just might, be a spy.

Experiment with Other Inks

Messages written using the following will become visible under a UV light:

- lemon juice
- vinegar
- diluted laundry detergent (with optical brighteners)

Here's how to keep your communication hidden:

YOU WILL NEED

- UV light source
- 5 glow vitamin tablets (see page 37) dissolved in 3 tbsp of vinegar
- Cocktail stick
- Cotton wool

1. Wrap a little cotton wool tightly around the end of the cocktail stick.
2. Dip the bud in the glow mixture and write your message. Use a UV light source to see what you're doing.
3. Pass your message to a friend and shine a UV light on the message to reveal your secrets.

Have a Candlelit Supper

Make mealtimes magical with mood-enhancing lighting

Eating by candlelight makes even beans on toast feel special. Indoor or out, just a few tealights scattered about a table create a fairytale atmosphere, but go big for a grander feel. The flickering glow creates a sense of intimacy and occasion, and it's a great way to cheer up a dreary winter teatime.

YOU WILL NEED

- Candles and secure candlesticks
- OR candle lamps
- OR tea lights and holders
- OR battery-powered versions
- Your supper

OPTIONAL

- Jars (keep a stack of interesting ones to hand – our favourite is an ancient cockle pot!)
- Fairy lights

- Kids love helping to light the candles – use a long lighter to keep things safer.
- Create drama by placing lights on different levels – use candlesticks of different heights, or place tea lights in holders or jars on sturdy objects such as pretty boxes or upturned vases.
- Scour charity shops for old candelabras, candlesticks or lanterns. Designate some as outdoor furniture and keep them in your garden for al fresco dining. The more rust and patina, the better.
- Place mirrors on your table to reflect the light.
- Scatter battery-powered fairy lights around the table.
- Customise jam jars using paints, ribbons or bunches of flowers tied around the rims and use them as candle-holders.
- Use unusual objects to hold your candles – hollowed-out squashes, tin cans, or pieces of driftwood all work well.

Stay Safe!

- **Stick to battery-powered candles or tealights if there are really small children about.**
- **Keep flammable table decorations well away from naked flames.**
- **Avoid reaching across the table if there are lots of candles ablaze – serve food well away from the flames.**

Games After Dark

Play Games Indoors

Perfect for parties, or to liven up a dull evening

Make winter evenings something to look forward to with a stash of play-in-the-dark games up your sleeve.

In the Spotlight

YOU WILL NEED

- A torch
- At least four players

1. Sit or stand in a circle with one player in the centre holding the torch.
2. Turn off the lights and turn on the torch.
3. The player in the centre of the circle spins round and shines the torch on another player.
4. That player should do a dance, sing, pull a face or tell a joke.
5. Alternatively, the first player starts a story, then each person in the spotlight adds another sentence.

Gallery of Statues

YOU WILL NEED

- A torch
- At least four players

1. Designate two players as the 'tour guide' and the 'tourist' then ask them to leave the room.
2. Turn off the lights.
3. The remaining players become 'statues' by getting into funny positions and freezing.
4. After 15 seconds the tour guide shouts 'the gallery is open', and enters the room, accompanied by the tourist.
5. The pair use their torch to explore the gallery, trying to make the statues laugh or move without touching them. Last statue standing wins!
6. For a variant, give each round a theme. You may like to try sports, animals or monsters.

Great fun for little kids!

Hide and Seek in the Dark

YOU WILL NEED

- A torch for each player
- At least three players

1. Give each player a torch and designate one player as the seeker.
2. Turn off the lights.
3. While the seeker counts to a hundred, the others go and hide, turning off their torches when they've found a good place.
4. The seeker uses their torch to look for the hidden players and as each player is found, they return to a designated point. Last to be found wins!
5. A variant is Sardines in the Dark. One player hides, then the others try to find them. When they find the hider, they join them. Last to find the group is the loser!

Good for slightly older kids

Murder in the Dark

YOU WILL NEED

- At least four players
- A completely dark room
- A pack of cards

1. Select as many playing cards as there are players. Make sure there's one Ace and one King in your selection.
2. Give each player a card. If you draw the Ace, you are the murderer. If you draw the King, you are the detective. Don't reveal your identities.
3. Turn off the lights.
4. The murderer 'kills' his victims by tapping them on the head. If you are 'killed' you lie down.
5. As soon as someone discovers a 'dead' person, they shout 'murder in the dark!'.
6. The lights are immediately turned on, and everyone assembles.
7. The detective comes forward. He may ask all the players questions – where they were, what they saw, and who they suspect.
8. If the detective is ready, they may accuse the murderer. If not, the lights are switched off. 'Dead' people should sit out, and another round is played.
9. The game ends when the detective guesses the murderer correctly, everyone is killed, or the detective is killed. If the detective makes a wrong accusation, the killer wins.

Fun for older kids and teenagers

Stay Safe!

- **Make sure your play area is safe before you start – nothing to trip over, no trailing wires, nothing sharp on the floor.**
- **Set boundaries before the games begin – no hitting, no inappropriate touching.**
- **Designate a 'safe word' – if anyone is scared, they can use it and the lights will be switched on.**

Torch Games

Perfect for camping trips, night-time park expeditions or when you've got a spare ten minutes in the garden after tea time

Torch Tag

You'll need a group of people and a big, dark space with plenty of places to hide. Choose someone or draw lots to be the seeker and designate an area as your 'jail'. The seeker is given the torch, closes their eyes and counts to 100 while the rest of the players hide. The seeker then switches on their torch and goes to find the others – you're tagged if the beam hits you and your name is called at the same time. If you've been tagged, you go to the jail. Last person to be tagged is the winner!

Variation: Try rolling torch tag. There's no jail; you just hand the torch to the person you've tagged, and they become the seeker.

Follow the Spot

An easy game for little kids. Simply use the torch to shine a spot around your dark space and when the spot stops, the first person to jump on it is the winner.

Glow Worm Tag

Choose someone to be the glow worm. They then run for 60 seconds then turn and flash the torch twice. Now everyone else can run and try to catch them. The glow worm must flash the torch twice every 60 seconds, and whoever catches them becomes the next glow worm.

The Hunt

This needs a little preparation (set it up in the day to make it easier), but it can entertain a group for a fair while. It works best in places your treasures won't be disturbed.

YOU WILL NEED

- Multiples of small objects such as wooden bricks, toy soldiers, marbles, painted rocks
- A torch for each player
- A bag or bucket for each player
- Timer *(optional)*

1. Hide your chosen objects. Decide on a time limit for the game and start the timer, or start counting.
2. Send out the players to hunt for the objects with their torches (you'll could give them clues about what they're searching for).
3. When time's up, shout for everyone to return. The player who has found the most objects is the winner.

Stay safe!

- **Always scope out your play space in the daylight.**
- **Look for trip hazards such as holes or roots, overhanging branches, or anything you could run into.**
- **Agree a word to shout if anyone is feeling scared or overwhelmed and stop the game immediately.**
- **Don't play any of these games near open fires or roads.**

Fluorolympics

Extend outdoor play time into dusk and beyond

Most existing garden games sets can be turned into glowing games. Or make your own – a bowling set using old plastic bottles weighted with sand or water, or ring toss set using weighted plastic bottles and thick-rope-and-duct-tape rings. These games are ideal to take camping or onto the beach.

YOU WILL NEED

- Fluorescent paints *(see page 71)*
- Fluorescent tape or dot stickers
- Garden game sets such as boules, giant Jenga, ring toss, bowling set
- UV light source *(optional)*

1. Paint your garden games sets using fluorescent paint. Stick to one colour or swirl and mix your paints or create patterns. Spots work really well, or you may prefer something more abstract.
2. Use tape or dot stickers to pattern up your pieces. We used tape to create bands on our bowling pins.
3. Use the tape to mark out play areas or lines if needed.
4. Take out your set and play into the night. Shining a UV light source onto your sets as you play gives even more get-up-and-glow but is not essential.

TELL ME ABOUT: Fluorescence

Fluorescent or 'neon' colours don't glow in the dark unless you shine a UV light on them.
However, in the day time, and particularly at dusk, they pop. Why? Because they appear to emit more light than they absorb. Regular colours take their appearance from the wavelength of light they reflect; our pink pins absorb every colour in the light spectrum apart from pink. At one end of the spectrum, just beyond violet, sits ultraviolet, or UV light. Fluorescent dyes absorb and convert UV light and other colours low in the spectrum, emitting them as visible light.

Try using glow-in-the-dark paint
or stickers, charge them up then
play your games on even the
pitchiest, blackest night!

Horror Feely Game

Raid your kitchen cupboards and create your own stomach-turning squeezables!

Great for Halloween or any dark night you fancy spooking up, this game turns everyday objects into sensory horrors that will have your friends and family screaming for more.

YOU WILL NEED

- Small cardboard boxes, drawstring bags or bowls
- Selection of items from the list below
- Willing participants
- Blindfolds *(optional)*

TO SET UP

Get together a selection of cardboard boxes, cut a hole in each, and then tape a fabric flap over that hole to stop peeping.

Alternatively, use drawstring cloth bags or simply place the items below in bowls and blindfold your participants. Now create your finger-frightening, touchy-feely scares!

TO PLAY

1. Sit around a table or in a circle on the floor.
2. Dim the lights and use battery-powered candles to create a suitably chilling atmosphere.
3. Set the scene by telling the tale of a mad scientist or witch who has left behind boxes filled with leftovers, explain that it's the players' job to feel their way through each unimaginable terror.
4. Describe each body part with grim glee. Pass each box, bag or bowl around and ask each player to feel inside. Try not to scream!

Sensory Squeezes to Try

Peeled grapes feel like…	eyeballs
Cold spaghetti feels like…	veins
A tinned tomato feels like…	a heart
Cold soba noodles feel like…	worms
Hotdogs feel like…	fingers
Wire wool pads feel like…	witch's hair
Dried apricots feel like…	ears
Almonds feel like…	fingernails
Cooked cauliflower feels like…	brains

Glow-in-the-Dark Tabletop Games

Playing board and tabletop games in the dark adds a whole layer of extra fun

You'll need to be nimble-fingered, mindful of space and depth, and cunning to win when you're playing with the lights out.

Why not customise some old favourites and give them a new lease of life (they'll still be fine to play in the day, too).

You can also pick up cheap second-hand sets at charity shops or boot fairs to experiment with.

YOU WILL NEED

- UV light source
- UV paint
  (or make your own – see right for details)
- Brushes
- Games

TO PLAY

1. Work out how to paint your game – e.g. can all the pieces be the same colour, or do some need to be different colours to represent different players?
2. Paint the pieces and leave to dry. Some pieces might need two layers to look really good under UV lights.
3. Invite a few friends over, turn out the lights and play in the dark!

Home-Made Glowing Paint

You can make your own blacklight-glowing paint from ingredients already in your home.

YOU WILL NEED

- Liquid washing detergent *(look for a brand containing optical brighteners)*
- Water-based paint
- Bowl

1. Pour a cup of paint into the bowl and add 1 tbsp of washing detergent.
2. Mix to combine using a wooden spoon.
3. Use as you would normal paint, and use a UV light source to make it glow. Experiment with brands and proportions to find the paint with the most get-up-and-glow!

Dance the Limbo

How low can you go?

This traditional Trinidadian funeral dance was popularised globally in the 1950s by the tourists who visited the island and in the UK by the Caribbean passengers who came to the country on the *Empire Windrush*. It became a favourite party game and inspired a series of hit records, including Chubby Checker's 'Limbo Rock', released in 1962. The song included the phrase 'How low can you go?', which is still used as a chant during the game. Today, it's still a bedrock of Trinidadian culture – as well as being danced for tourists it's also played by children in the streets.

Of course, you can limbo dance in the day, but it's more exciting at night. Many displays of limbo in Trinidad use a pole that's been set alight, with dancers having to wiggle bravely under the flaming rod, but it's safer to use fairy lights or neon wire to illuminate your limbo pole.

YOU WILL NEED

- A pole – *a broom handle or a straight stick will do. It needs to be as thick as your thumb and at least 1.5 metres (5 feet) long*
- Fluorescent stickers or tape
- Battery-powered fairy lights or electroluminescent wire
- String or rubber band
- Music
- At least two people

1. Decorate your pole with fluorescent tape or stickers. Wind the fairy lights or electroluminescent wire along its length, attaching the battery to the pole with string or rubber bands. Turn on the fairy lights.
2. Start up the music – a search for 'limbo' on a music subscription service will throw up some excellent party tunes, or find my limbo party playlist at tiny.cc/limboparty. Alternatively, sing your own songs and clap along.
3. Get your players to line up and hold the pole horizontally at your tallest players' chest height – it's best to have two people doing this, but you can just about manage with one. If you're exceptionally well-prepared, or have weak arms, make stands to hold the pole!
4. Get your players to pass under the pole in turn. The players must always bend backwards and are eliminated if any part of their body (except their feet) touches the ground, or if any part of their body touches the bar.
5. Actively encouraged: doing a warm-up, show-off dance before passing beneath the bar. Ridiculous celebratory jigs after you manage to wiggle beneath!

Limbo Tips!

- Spread your feet a little – to the same width as your shoulders.
- Open your arms wide
- Take things slowly
- Keep your stomach flat (limbo dancers have incredible abdomen muscles!)
- Keep your head back until you're completely under the stick

Epic Outdoor Group Games

Make use of big spaces and play into the night

There's a thrill and energy to playing outdoors in the dark. Games feel wilder, more on-the-edge, and grown-up. These suggestions are perfect for playing with friends and family while camping, staying late at the beach, or wrapped up on a wintry night in the park.

Ghosts in the Graveyard

Get spooky with this chilling chase game.

YOU WILL NEED

- At least three players

1. Designate a base and determine the boundaries of the play area or 'graveyard'.
2. Choose one or, if the players are younger, two players to be ghosts. The ghosts run off and hide (together if they're in a pair).
3. The remaining players count to a hundred, then roam the 'graveyard' in search of the ghosts.
4. The player who spots a ghost first shouts, 'ghosts in the graveyard!'
5. Now the ghost(s) are allowed to chase the players until a fleeing seeker is tagged – they also then become a ghost for the next round, so the number of ghosts will increase each round.
6. The winner is the last player to be left alive.

Capture the Flag

This all-time classic needs as big an area as possible in which to play. Ideally, there will be plenty of places to hide. Dunes, parks or campgrounds are perfect. However, you can play a smaller version in a garden. The rules may seem complicated, but once you get going it's very easy. Feel free to add your own variations to mix things up.

YOU WILL NEED

- Two 'flags' of different colours or designs – these could be actual flags, scarves, or even old T-shirts
- At least 6 players (the more the better)
- A head torch or torch for every player

1. Mark out the playing area using torches, lamps, cones, T-shirts or trees. You'll need a boundary and a strong dividing line down the centre of the playing area. You can use naturally occurring lines – between two trees, a point on a building, or even a stream.
2. Designate an area or object in the centre of each team's territory as a 'jail', and an area or object as a 'home base'.
3. Make sure everyone knows exactly where the boundaries are.
4. Split the group into two even teams.
5. Each team has two 'hiders' who should run off and conceal their flag somewhere in their own territory. The remaining players cannot sneak a peek at where the flags are being hidden. The flags should be easy to grab, not buried, but can be hidden in fairly tricky places.
6. The hiders return, and the game begins.
7. If you're running, you must turn on your head torch or torch. If you're walking, you can be stealthy and turn it off.
8. Once players enter enemy territory, their opponents will try to 'tag' (lightly touch) them. If a player is tagged, they must go to jail in their enemy territory. Players may keep their torches on in jail in order to be seen by their team-mates.
9. Players can be set free if one of their team comes to rescue them with a touch. They must return to their own home base before being allowed to return to their opponents'. However, this walk is 'free' – they cannot be tagged.
10. If a player carrying the flag is 'tagged', they must drop their prize and go straight to jail.
11. The winning team are the group who return the flag to their own home base.

Cops and Robbers

This game is more suited to older kids and teens and requires a huge playing area – two to five miles is perfect! This can stretch late into the night, so make sure everyone has supplies. Pair players up and choose safe areas with light traffic to play in.

YOU WILL NEED

- At least six players
- A torch for every player

1. Designate one spot as the bank, another (at least a mile away) as the safe house – a real house or tent is perfect.
2. The game starts at the bank. Split the teams equally into cops and robbers and give each player a torch for safety. The robbers have just pulled off the heist of the century, and need to get the loot back to the safe house.
3. The robbers are given a one minute head start, and start to make their way back to the safe house. They can hide, sneak around back alleys, and take less obvious routes in order to get there without being caught.
4. The cops follow. If they touch a robber, that robber is caught, and must make their way to the safe house as a prisoner.
5. Vary things up by allowing cops to ride bikes, or all play on bikes.
6. Cops and Robbers can last for hours! Put a time limit on the game, and set a time when everyone needs to be at the safe house.

Play Safe!

- **Be aware of trip hazards.**
- **Never play anywhere near traffic.**
- **Smaller kids should pair up with a grown-up.**
- **Play at dusk rather than in the pitch black to make things safer.**
- **Have a safe word – if anyone uses it, stop the game and turn on your torches.**

Nature After Dark

Smell Night-Blooming Flowers

Have a nose around your garden or park

The heady aroma of night-blooming flowers is one of nature's more secret gifts. Take a wander through your local park, or anywhere with planted beds, in spring or summer and breathe deeply as you pass the beds and blossom-bedecked trees to sniff out the sweetest smells.

Petal Power

Honeysuckle

This sweet-smelling plant is found trained up walls, large honeysuckle bushes are often a hive of natural activity. Birds nest in them, and butterflies and moths are attracted to the blooms. Most fragrant of all is the native common honeysuckle, bright with pink and yellow flowers. Try picking a single petal and sucking the honey from the base.

Tobacco Plant

No, it doesn't smell of cigarettes – thank goodness. The long, elegant trumpets of this tall plant is intensely sweet. The British Victorians planted it along wooded paths to create intoxicating places to wander with their sweethearts at dusk. Don't be tempted to touch it, however, it's poisonous and sticky, so sticky, in fact, that it sometimes traps mosquitoes on its leaves and stems.

Star Jasmine

The five-fingered, pretty white flowers of this climber are often found trailing from walls. The scent of this plant has long been bottled – ancient Egyptians and Greeks believed it was a cure for stress and anxiety.

Evening Primrose

Not only can you smell this bright yellow plant, but, if you're very lucky, and very quiet, you can hear it. As the buds burst at twilight, they pop! Its vanilla-like scent lays heavily across parks, dunes and waste ground. It's long been prized for its medicinal qualities – early herbalists called it the King's Cure-All – and is still used to make curative oil.

Sweet Rocket

You'll find these pale lilac or white flowers in borders in early summer. Bees love the sweet, violet scent of this plant that is strongest on warm evenings. Grow them in your garden and use the petals to decorate cakes and flowers – they're edible.

TELL ME ABOUT: Night Scent

Why do these plants smell at their most sweet after the sun goes down? It's usually because they're pollinated (fertilized to create seeds) by night-flying insects – moths and mosquitos, even some beetles. The flowers are sensitive to changes in light, unfurling when these creatures emerge in the evening. Their petals are often light-hued, all the better to attract these nocturnal helpers.

Listen to Nature

Shhhhhh! Take a wild soundbath

When did you last sit still? Really still? And do nothing but listen? We so rarely take time out to concentrate on one sense, but as night falls, and it's harder to see, our hearing naturally becomes more sensitive. What will you hear?

YOU WILL NEED

* Your ears

OPTIONAL

* Something waterproof to sit on – a groundsheet or plastic bag
* Cushion
* Torch *(one with a red light option or covered with red plastic wrap or cellophane is best)*
* Sound recording equipment *(a phone will do, but the screen will spoil your night vision)*

1. First, find your spot. Perhaps you'll go listening in the woods, in a park, by a lake or on a beach. Maybe you'll just head into your back garden, or even sit by an open window.
2. Sit somewhere comfortable – use a cushion if it helps. Practice sitting still; get comfortable, relax and concentrate on breathing in and out. Close your eyes and get your night listening going (see page 137).
3. Be quiet (no talking or whispering if possible) and listen hard. You may hear the swoosh of wind in leaves or grasses, or distant traffic, but let your ears do their thing and soon you'll pick up more alien sounds; an owl hooting, something scuttling through the undergrowth, crickets, or the yowls and screams of foxes and badgers. You may hear frogs singing to each other or, if you're lucky, the boooooom of a bittern or the sweet song of a nightingale.
4. You might want to try to identify all the calls, trills and rustles, but it's also rather nice to freestyle and let them wash over you like a natural symphony. It's a way to feel connected to nature; feel the earth beneath you and the sounds of the wild world in your ears.
5. Want to hear even more? Cup your hands behind your ears for super-powered listening action.
6. Did you bring a recorder? Why not let it run and pick up the sounds that surround you. You can listen back to it later if you'd like to work out what creatures you heard, or even use it as a natural, home-made sleep aid.

Mother Nature's Microphone

Small children may not be able to sit still for long, so make this pretend recorder to help them focus on listening.

YOU WILL NEED

* A stick
* Something fluffy – tufts of sheep's wool or cotton wool
* String

1. Tie the fluffy thing to the end of the stick with string.
2. Take your 'recorder' out after dark.
3. Listen for sounds, pretend to record them, then 'play' them back using your voices. You might find yourself impersonating an owl, making papery leaf noises, or miaowing like next-door's cat.

Go on a Night-Time Safari

Spot the shyest of creatures after dark

Spotting animals at night is a whole lot more exciting than in the day. Pet cats and dogs are tucked up, many birds are asleep, but the nocturnal (see page 87) animals come out to play. Perhaps it's because their lives are the total opposite of our diurnal (see page 87) existences that we are so fascinated by the creatures of the night.

YOU WILL NEED

* Torch *with a red light option or covered with red plastic wrap or cellophane*
* Something waterproof to sit on – a groundsheet or old plastic bag.
* Warm, waterproof clothing

OPTIONAL

* Notebook
* Pen
* Snacks
* Warm drink
* Cushion

Stay Safe!

- **Take an adult with you, or make sure they know exactly where you are and when you intend to be back.**
- **Never scare, chase or touch an animal.**
- **Stay clear of any areas you know are home to more dangerous animals and make sure you know what to do in the event of spotting one.**

You can spot wild animals anywhere. No matter how small, your back garden or balcony will be home to a population of creatures (we have badgers, foxes and lots of woodlice in ours). However, heading out to a park, a wood or the countryside will increase your chances of seeing more. Find a dry, comfortable spot and settle down, and let your night vision kick in (see page 124).

Nocturnal animals are good at hiding, so it's important to learn how to be quiet yourself. Learn to stalk – walk with slightly bent knees, arms at your sides. As you creep along, try rolling your feet from their sides rather than stomping noisily, and be aware of sticks underfoot that might snap or make noise. Feel your way. Don't wear scent – some animals will be able to smell you and will stay out of the way.

Look up to see bats and nocturnal birds. Look down to see beetles, spiders, mice and small creatures. Peer behind bushes. Listen hard – see Listen to Nature (page 82) for more inspiration.

Note down your sightings in your book. You may like to sketch the creatures you see, or write down how they sound, what movements they make, or where you saw them. If you spot footprints, why not record them too?

Go for the Glow

If you're really lucky, you may spot glow-worms (AKA fireflies or lightning bugs). Not actually worms, but beetles, these nocturnal creatures use their natural luminosity to attract mates. Adult female worms are bioluminescent – a molecule called luciferin combines with oxygen to create oxyluciferin. This then reacts with an enzyme, luciferase, to illuminate a large organ at the end of their abdomens. In the UK, you're most likely to see them in June or July, often in limestone-heavy areas where there is a good supply of small snails. Check online to find sightings maps or a glow-worm walk near you.

Where to Explore

- **Ponds, rivers and streams are home to hundreds of different creatures.** Listen for the summer night songs of water boatmen and watch for them skating across the surface of the water. Use your red torch to search for frogs, their heads peeping up from below the surface. In spring you might see (and hear) them mating and laying thick, gloopy frogspawn. You may spot toads, newts, voles and bats here too (see page 94).
- **Woods, fields and forests are home to bigger creatures.** Look for deer, rabbits, hedgehogs, badgers, weasels, foxes and stoats. Sit quietly and listen for them rustling through the undergrowth and calling to each other. Use clues to help you find a good spot to see animals; hairs caught on barbed wire, footprints, and sniff for the strong scent of badgers and foxes. Find a spot near a sett, a burrow or rabbit warren around dusk, sit and wait. Stay long enough and quiet enough and you should spot your quarry. Try not to get too excited when you spot something – you don't want to scare them away.
- **Towns are surprisingly full of wildlife.** Take an urban walk to spot foxes under street lamps, rats and mice scuttling along pavements, hedgehogs on their trails across garden lawns and, if you're very lucky, badgers bumbling through bushes.
- **Even tiny gardens and balconies have a creature population.** Look for the slimy, shiny trail of earthworms, slugs and snails and follow them to track down your mini beast. In summer, these animals prefer to come out at night, when it's cool. Or find spiders tucked in corners. Lift up pots or rocks to see what's underneath; you might find beetles or bugs. Airborne creatures are everywhere – spot birds, moths (see page 88) and flies. Light a lamp and see what you attract.

Make a Track Station

No-one can stay up all night watching for creatures. Record the footprints of lurking animals with a baited board that records visitors.

YOU WILL NEED

* A large piece of plywood board or cardboard, or big play tray
* Flour, cornflour, play sand, or soft dirt
* Flat dish or block of wood
* Bait such as peanut butter, cereal or bird food
* Spoon

1. Find a flat spot at the edge of an open area, near shrubs, trees, or grasses.
2. Place the board flat on the ground and smooth your powdery material over it – it should be about 1 cm (½ in) deep.
3. Put the flat dish or block of wood in the middle of the board and spoon some of your bait on top.
4. Leave it overnight.
5. When you return in the morning, you should see tracks and marks. See if you can identify any of them.

What it means

NOCTURNAL animals come out to feed and go about their business at night.

DIURNAL animals emerge in the day to potter about.

CREPUSCULAR (lovely word) animals are active in the hours during twilight – at dawn and/or dusk.

Hunt for Moths

Attract mysterious winged creatures to your garden

Moths are the butterfly's gothic cousins, mostly coming in sludgy shades of brown, green or grey – although there are many colourful exceptions. Many people think of them as creepy creatures, or pests whose larvae like to nibble on your best jumpers, but moths are a fascinating and beautiful species.

Look out for the spectacular Emperor, its back emblazoned with four 'eyes', or the Mottled Beauty, whose cleverly mimicking colouring adapts to the particular region in which it lives. If you're very lucky, you may spot the Death's Head Hawk Moth, which squeals and squeaks and has a clear skull marking on its back – beware though, its appearance is believed to be a bad omen.

The easiest way to find moths is to wander around your garden, a park, or a wood with a torch. They're attracted to blossom and strongly scented flowers such as buddleia or tobacco (see page 80); finding these plants will increase your chances of seeing plenty of the winged creatures. However, the easiest way to lure them from the darkness is to make a light trap.

YOU WILL NEED

* Single bedsheet
* Bright torch
* Rope or washing line
* Pegs

OPTIONAL

* Camera
* Notepad

1. Find an area with little light pollution and with as many plants nearby as possible.
2. Peg the sheet onto a washing line or sling a rope between two trees and suspend it.
3. Shine the torch directly onto the sheet. The moths will be attracted to the light beam and land on the sheet. Take a close look, and photograph or sketch them if you'd like to identify them later. Don't touch the moths though, as this can damage their delicate wings.

Moth Attraction

To lure moths to your garden, mix a little rum (ask an adult to help with this!) and treacle and paint it on tree bark to draw these fragile creatures close.

If it's cold or wet outside, you may be able to attract moths to your house. Place a lamp near a closed window, turn all your other house lights off, and watch the flying creatures settle on the glass outside.

Climb Trees by Torchlight
Head upwards as the sun sets

Scrambling into the branches of trees is pretty exciting by day, but at night it's an even greater adventure. Everything is amplified, from the rustle of the leaves, to the distance to the ground. You'll not only feel closer to the sky, but closer to nature too.

YOU WILL NEED

* Warm clothing with long sleeves
* Head torches

Find Your Climbing Place

For maximum impact, select a site with spectacular views, whether they're across the countryside, the beach or the city. Even though you're only headed a few metres upwards, a vista looks extra special when you witness it through branches or leaves. Make sure there are plenty of great climbing trees around – healthy, strong and with branches close to the ground. If anyone is nervous about heights, they might prefer to scramble along a fallen tree.

Make Sure it's Safe

For safety, scope out your route in the daytime; make sure there are no loose or rotten branches or unexpected twigs that might poke you in the face. Will you all climb one tree? Or perhaps each member of the group will select one. Work this out before it gets dark. If you have really young kids, illuminate the way up the tree by stringing battery-powered fairy lights along their route.

Try Climbing as the Sun Sets

You'll catch the best light and see the landscape at its most beautiful. Always, always take a head torch with you, secured before you start your climb, whether it's light or not; you can turn it off when you reach the top. Keep shimmying upwards from branch to branch, feeling your way, until you find a perch, and then...

- **Gaze...** at the night-time vista.
- **Listen...** for scuttling animals, distant traffic, or church bells.
- **Hug...** the tree and imagine what it has witnessed in the years it's been alive.
- **Grin...** at the magic.

Then make your way back down. Careful!

Fallen in love with tree climbing? Why not try sleeping overnight in a camping hammock? Half-tent, half-hammock, these cocoons hang high in the trees and make for a comfortable den.

Climb Safely!

- **Be extra sure that each branch you reach to or step on will hold your weight.**
- **Always carry a head-mounted torch with you.**
- **Wear strong, well-fitting shoes.**
- **Stop if you get nervous.**

Take a Moonlit Beach Walk

Seaside creatures come alive and glow after dark

As the sun dips below the horizon beaches take on a different personality. Empty and often barely lit, they become thrilling places to explore. For brighter ambient light, pick a night when there's a full moon. Not only will you be able to see better, but tides are lower, so there's more nature to spot.

YOU WILL NEED

* Torches and spare batteries
* Tough, waterproof footwear (beach shoes are great) so you can scramble over rocks

OPTIONAL

* UV light
* Whistles
* Clear buckets or containers to allow you to view your finds from all angles
* Magnifying glass
* A flask of hot drink

Listen to the Birds

Not all marine birds sleep at night. Keep your ears open for the sound of waders, piping oystercatchers and the distinctive coo-ee shriek of the curlew.

1. **Make sure everyone has a torch, and, if it's really dark, a whistle, in order to stay safe.** A walk along the shoreline, picking out waves with your light is so much more dramatic in the dark, and you may see the silver flash of fish leaping from the water.
2. **Many species of marine life are at their ease in the dark** – fish come into the shallows and crabs go on the hunt.
3. **Always busy places, at night rock pools are like rush-hour city centres.** Shine your conventional torch on one and you'll spot shrimps and prawns, fish and hermit crabs ambling across the floor. Use your hands to pick them up for a closer look – nets destroy the delicate habitat. Put the creatures and plants in a transparent bucket and look at them from below for a different perspective.
4. **If you've brought your UV light, now is the time to turn off your conventional torches and switch it on.** If you're lucky, you'll get an impressive, marine creature light show.

Rockpool Radiance

Snakelocks anemones glow bright green under UV rays, their fronds waving like Medusa's mane and lighting up the rockpool. It's a truly impressive phenomenon. Gem anenomes glow too, in green with starbursts of orange, while daisy varieties shine in dull red. Try training your UV light on seaweeds; some varieties are bright red, pink and orange, while grey topshells have a pink tip to their carapaces. Certain species of crab have a blue tinge.

Shore Show

Marine biologists are not sure why some sea creatures fluoresce (glow under UV) – perhaps because of a symbiotic algae (zooxanthella) that grows in them, or maybe as sun protection. Whatever the reason, the phenomenon is stunning.

Go Bat Spotting

Who's flapping overhead?

You'll find bats in most areas – once you start looking for them you'll realise you're surrounded by these flying mammals. All you need to do to is head into a garden, a park, or the countryside and gaze upwards.

Bats are at their most active during the summer, on warm, still, dry nights, when their on-board echolocation system works best. Choose a dark area away from streetlights if you can. You'll often have more luck spotting them near a body of fresh water, because they swoop to take drinks from ponds, lakes and canals – stand on a bridge or somewhere else with a good view of the water.

Taking Flight

The best time to go bat spotting is around sunset, and for an hour afterwards. Different species come out at different times of the evening and have distinctive flight patterns – some may swoop in straight lines, others zig-zag. You'll see them dip down as they scoop up flying insects for an in-flight snack.

Listen In

If you get really serious about batspotting, why not borrow or buy a special detector? They work by converting the high, ultrasonic noises that bats make into a sound low enough for the human ear to hear. This enables us to hear what kind of bat is flying overhead, as each species has a distinct call. You'll find detectors online – bat protection charity sites recommend good brands – or join a local bat group to borrow theirs and learn more about these fascinating creatures.

DID YOU KNOW?

- One in five mammal species worldwide is a bat.
- There are over 1,200 species, with more being discovered all the time.

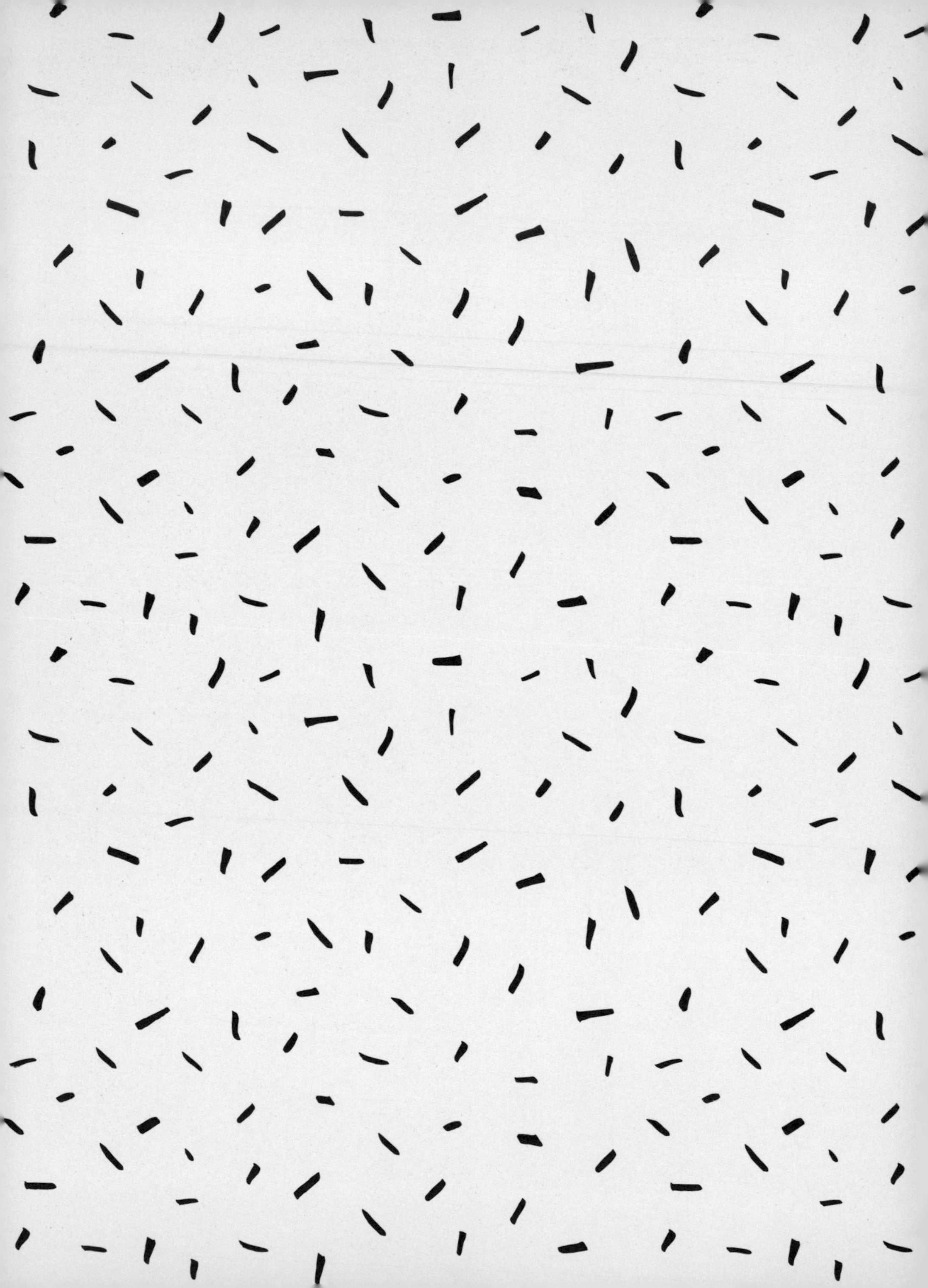

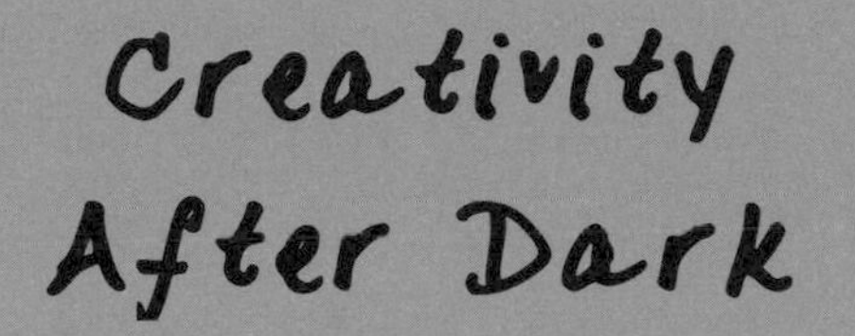

Creativity After Dark

Night-Time Photography

Get snap happy and use the darkness creatively

Night-time camera fun might need a little more thought than day-time snaps, but you can get atmospheric results and, by experimenting, create visual magic.

Shoot-Spiration

- Play with silhouettes. Stand against the sinking sun or a big light source and create shapes with your bodies or other objects.
- Shine coloured torches into the camera to create abstract light paintings.
- Take pictures of your neighbourhood. Shoot people standing under lamp posts in pools of light, or the sun setting over roofs. Skylines and cityscapes also look stunning in the dark.
- Head to after-dark carnivals, funfairs or firework displays and document them. Turn your camera on the audience's faces for a fresh take.
- Use reflections to create surreal effects; mirrored buildings, the sea, ice or lakes.
- Secure your camera on a tripod and experiment with long shutter speeds. Using a digital SLR, adjust your settings to a small aperture and low ISO and set the shutter speed to at least five seconds. Get your models to wave torches or sparklers and capture the patterns they create. Or shoot a road scene, and watch the car headlights in your picture become streaks.
- Experiment with ICM (intentional camera movement) – set a longish exposure (up to 1 second) and move your camera as you shoot to create coloured, abstract blurs.

Night photography five golden rules *

1. **Turn off the flash**. It results in harsh photos, filled with hard shadows. Use alternative light sources such as torches, lamps, candles or the setting sun. See how your pictures change as you alter the direction of your light source or use coloured cellophane or sweet wrappers to change its colour.
2. **Use a tripod.** Essential to avoid camera shake, or you could improvise your own by propping up your camera on something solid. Hold your breath as you press the button to keep things even more steady.
3. **Experiment.** Digital cameras give you lots of chances to get your shot just right. Try out different settings, shoot the same shot from down low or up high, or get your model to act out different emotions.
4. **Practice.** Take pictures all the time, whether you're at a special event, on holiday or just walking home from school. Make it a habit and your pictures will improve.
5. **Edit!** Use an app or program to make your pictures even more beautiful. Bump up the contrast, exposure, or add filters or themes.

* But remember, rules are there to be broken!

Put on a Shadow Puppet Show

Tell tales using silhouettes

The art of shadow puppetry originated in ancient China thousands of years ago. Complex stories were played out using intricately designed figures and cleverly engineered props, all accompanied by live music. You may have to scale things back a little, but the timeless, minimal nature of shadow puppetry still brings the drama. Why not try creating your own plays?

Make Shadow Puppets

YOU WILL NEED

- Card
- Pencils
- Sticky tape
- Scissors
- Tape
- Wooden skewers, balsa sticks or chopsticks
- Light source such as a lamp or torch

OPTIONAL

- Split pins *(also known as paper fasteners or brads)*

1. Using a pencil, draw your puppet designs onto the card – use your imagination, trace outlines from colouring books, or print out templates found online. You might create characters, animals, houses, cars, carriages, cups or monsters. Think about how your story will work, and what characters and props you'll need to make. Your designs must to be in one unbroken shape unless you're using split pins to give them joints.
2. Cut out your puppet shapes.
3. Stick the skewers or sticks to the back of the puppet shapes using sticky tape.
4. Get creative and use split pins to add joints to your puppets – you'll need more skewers or sticks to support limbs, legs, jaws and arms.
5. Turn on your light source and use your puppets to cast shadows on the wall or in your theatre.

Create the Theatre

You've got your puppets, now write a script. A show based on traditional fairy tales will keep things simple – have a rehearsal and gather an audience.

YOU WILL NEED

- A strong light source such as a lamp or torch
- A large cardboard box
- A large sheet of white tissue paper or greaseproof paper
- Sticky tape
- Scissors
- Stools
- Marker pen *(optional)*

1. Cut the bottom out of the box and cut off one of the long flaps on the top of the box – this will be the top side of your theatre.
2. Stretch the tissue paper across the hole in the bottom of the box, securing it with tape. It's easiest to do this on the outside of the box, but neater if you can manage to do it inside.
3. Put the box onto a stool, place the light source on another stool behind the box and turn it on.
4. Puppeteers should sit beneath the box, their puppets between the light source and the screen.

Lightbox Fun

Add shine to playtimes

At its most simple, a lightbox is just that: a light in a box, softened. Artists use them to trace outlines, photographers use them to look closely at negatives and prints, and craft fans use them for calligraphy or stenciling. We can use them for all kinds of fun things. And they're a LOT easier to make than you might think.

YOU WILL NEED

- A clear plastic box or other clear container
- Battery-powered, white fairy lights
- Aluminum foil
- Tissue paper

1. Line the box with aluminum foil
2. Place the lights into the box – if they're battery-powered, great. If not, run the cord out of the top of the container, or cut a hole in the edge of the lid to allow the cord to pass through.
3. Stick a sheet of tissue paper to the inside of the lid
4. Put the lid on top of the box
5. Play!

Bright Ideas

Some fun things to try with your lightbox:

- Find leaves to examine on your lightbox – turn it on to see their veins.
- Sprinkle sand and shells onto your lightbox and make patterns that glow.
- Use it to trace characters from your favourite comics.
- Place coloured sweet wrappers, transparent building blocks or game pieces on it and use them to create pictures.
- Do a jigsaw in the dark.
- Roll marbles around the lid and watch them light up.
- Use translucent, differently hued shapes to experiment with colour mixing.
- Use felt-tips to draw on transparent plastic to create stained glass effects.
- Download printable x-rays and pretend you're at a hospital or vets.
- Make translucent numbers and letters to make spelling and maths more fun.

Twinkling Tin Can Lanterns

Create a cosy atmosphere with these simple hand-made lights

Tin can lanterns work indoors and outside, bringing magic and glow to dark corners and casting dancing shadows on walls and across gardens. They're simple enough for kids to make, although supervision is needed with the hammer and nails. Make them in bulk for a dramatic effect – use to line garden paths, hang from branches, or as table decorations.

They make lovely presents; personalise them with names or with single letters, arranged to spell out messages.

YOU WILL NEED

- Clean tin cans of any size *(peel-top designs are safest for little hands)*
- Hammer
- Nails
- Battery-powered tea lights
- Paper
- Rubber bands

OPTIONAL

- Wire
- Sand
- Small beanbags
- Spray paint
- Marker pens

As the cans rust and oxidise, they look even more striking – they redden and age beautifully. However, if you want yours to stay fresh looking, use spray paint to coat them. Matt blacks and greys look particularly elegant. Or try using marker pens to add colour.

1. **Fill the tin cans with water or wet sand and freeze for at least five hours.** This gives them some solidity and stops them from rolling about. If you want to keep them even more firmly in place, use small beanbags or cushions.
2. **Plan your patterns and draw them on your paper.** They should be relatively simple and small enough to fit on the can. Try hearts, flowers, letters, butterflies or stars. Once you get good at the technique, you can attempt more complex designs; game or cartoon characters and faces. Search online for templates and inspiration.
3. **Secure the paper designs** to the cans with rubber bands.
4. **Use the hammer and nails or awl to punch holes following your lines.** Try using different sizes of nails to create varied effects.
5. If you'd like a **handle** on your lantern, punch two holes near the can's top, and string some wire between them.
6. **Run your can under a warm tap** to encourage the ice to melt, then dry.
7. Place a **battery-powered tea light** into your can.

Make
Glowing Art
Masterpieces to
brighten nights

Be inspired by the growing number of outdoor winter art shows in our parks and city centres to create your own gorgeous, glimmering creations.

Glow painting

YOU WILL NEED

- Paper or card
- Glow-in-the-dark or fluorescent paint *(see page 71 for a quick and easy recipe to make your own)*
- UV light source *(if using fluoro paint)*

Keep things simple for really young children – a set of fluorescent or glow-in-the-dark paints and a sheet of paper under a UV light source can keep tiny artists absorbed. There's something magical about using glowing paints – pictures leap into life.

Why not hold your own Glow Art exhibition in your garden or street and invite friends to your opening night?

Glow Sculptures

YOU WILL NEED

- Glow in the dark paint or fluorescent paint
- UV light source

OPTIONAL

- Twigs, branches or driftwood
- Stones
- Shells
- String
- Recycled materials – *bottles, boxes, old technology, bike parts, containers, packaging*
- Glue gun
- Scissors or craft knife
- Fluorescent tape
- Throwies (see page 130)

- Collect natural materials to paint – stones, branches, shells or driftwood – and use your materials to create bigger pieces of art; mobiles, collages or sculptures that shine under UV light.
- Bigger kids can scale things up, try using long rolls of art paper, or work on flattened cardboard boxes.
- Find a wall or fence to practice graffiti (use wash-off paint).
- Paint the trees in your garden to create a space perfect for raving or to hold your own mini festival (see page 38).
- Go big and hang painted junk from trees to create an over-sized mobile.
- Use sound – bells, shakers or pieces of metal clanking to add drama to your art pieces.

Make a Teacup Candle

So chic, so easy!

Create a batch of these delicate lamps and light up a room or garden. We scour boot fairs and charity shops for vintage cups, vases, jam jars, and enamel mugs to make ours.

YOU WILL NEED

- Teacup (or vase or small pot) – *check carefully for cracks*
- Soy flakes or beeswax pellets
- White cotton string or pre-made wicks
- Scissors
- A pencil or skewer
- An old saucepan or bowl and saucepan
- Hot glue gun, duct tape or glue

OPTIONAL

- Essential oils – *pick a single scent, or blend your favourites together*
- Pliers

1. Measure the soy flakes or beeswax pellets. Fill the teacup with wax and empty it into the pan. Do this twice.
2. Cut a piece of white string 5 cm (2 in) longer than the height of the teacup.
3. Melt the wax gently in the saucepan over a low heat, or in a double boiler (a bowl floated in a pan full of boiling water).
4. Soak the string in the wax and pull it out using pliers or a fork. Let it cool on a plate or piece of tin foil. Take the wax off the heat.
5. Using a hot glue gun, a tiny piece of tape, or glue, stick the wick to the centre of the bottom of the teacup.
6. Roll the top of the wick around a pencil and gently pull it straight. Balance the pencil across the top of the teacup to ensure the wick stays standing straight up.
7. Slowly reheat the wax in the pan until it melts again.
8. Shake the essential oils into the teacup, if using. Be generous – we use 30–40 drops in our cups.
9. Gently pour the hot wax into the cup.
10. After 24 hours trim the wick to 1cm. Your candle is ready to use.

Glow from Head to Toe

Make your clothes shine and light up the dancefloor

Fluoro Feet

Customise your shoes with fluorescent pens and tape. Light-coloured canvas trainers work best. UV-reactive colours look wild in the day, but get even more intense as the sun starts to sink (find out why on page 66). Deck them out with fluorescent laces to give them extra zing. Shine your UV torch or light source on them at night for a truly far-out effect. Any designs look good – scribbles, polka dots, flowers, stripes or your name. This method also works well on light-coloured canvas hats, belts, or bags.

Gleaming Jeans

Stitch designs into your denim with glow-in-the-dark thread (available at big haberdashers or online). Follow the seams to create an outlined effect, or go freestyle and stitch designs across the fabric. Glow-in-the-dark threads can be charged in the sun or under a light source then shine brightly in the dark.

Lit Knits

Did you know you can buy fluorescent yarn? Use it to create friendship bracelets that glow as your relationship grows, to stitch slogans across bags, sweatshirts or belts, or, for a show-stopping look, knit the wool into jumpers. Hats and scarves made from the yarn not only look great, but also ensure you are more visible when walking or cycling in the dusk. Fluoro clothing equals road safety.

Luminous T-shirts

Fluorescent fabric paints take a simple T-shirt from plain to insane. Even unpainted white fabrics glow under UV lights, particularly if you wash them in powder with optical brighteners (see page 66). So choose lighter colours for maximum razzle. Then get your brushes out and get creative. We love flicking paint to make cool, dripping abstracts, or stenciling disco slogans across our chests. Don't forget the arms and back!

Electro Trickery

Take fashion to flashion with sewable LEDs and flexible electroluminescent wire. These battery-powered lights can be tacked onto clothing, or taped onto bags, their small controllers and power sources sewn into pockets or stuck in discreet places. No need to charge them up or shine UV lights on yourself; these shine and shine into the night. Try them tacked along shoe soles, marking out necklines, or spelling out names or words.

Make Ice Lanterns

Frosty lights for wintry nights

These magical lights are super-easy to make and look beautiful. If it's cold enough, make them outside. If not, your freezer is your friend. Use them to line an outdoor path in winter, to light-up a garden feast, or pop a selection along your garden wall at Diwali, Christmas or Hanukkah.

YOU WILL NEED

- Round plastic containers or tins of varying sizes
- Water
- A freezer or a cold, cold night
- Tea lights (real or LED)
- Sticky tape
- Weights, or a few small stones and rocks
- Twigs, berries, leaves, small shells or stones
- Scissors

Ice Magic

Why not freestyle? Try using small toys, game pieces or glittery bits to make your candle holders brighter and sillier. Experiment with food colouring in the water to create lanterns in rainbow shades.

1. Choose your plastic containers wisely. You'll need one larger, and one smaller (but still big enough to fit a tea light into comfortably).
2. Put a small amount of water into the larger plastic container –fill until it's about 5cm deep.
3. Put weights or rocks into the smaller container and place it into the first. It should float but not tip over. Carefully secure in place using sticky tape.
4. Cut the leaves and twigs if needed and poke the decorations between the two containers.
5. Add more water until it's about 5cm from the top.
6. Put the containers into your freezer or outside if it's cold enough and leave overnight.
7. Check the lantern has frozen solid, remove the tape and turn out. Run warm water over the outside of the container to help things get moving.
8. Put your lantern onto a plate to catch any drips, place a tea-light inside, and light it.

Create Silhouettes of your Family and Friends

Make like a Victorian and get outlined

Silhouettes made inexpensive keepsakes during the 19th century, before the rise in popularity of photography. Mothers would hang black-and-white cut-outs of their children on the wall or tuck them into the family bible. Really good silhouette artists such as the French-born Auguste Edouarte worked by cutting images directly using scissors. This method is a little easier, quicker and safer!

YOU WILL NEED

- Large sheet of white paper
- A4 sheet of black or coloured paper or thin card
- A4 sheet of white paper
- Strong light source such as a lamp, torch, or projector
- Sticky tape or tack
- Good scissors
- Soft pencil
- Glue
- Stool

OPTIONAL

- Frame

1. Tape or tack the large sheet of white paper to a wall.
2. Position your model on the stool.
3. Move the light source around until your model's shadow is thrown sharply onto the white sheet of paper. Their head and a little of their shoulders should be undistorted and small enough to fit a piece of A4 paper. Secure your light source or get someone else to hold it.
4. Ask your model to sit very still. Draw around the shadow's edge using a soft pencil, working as fast as you can.
5. Take down the large piece of paper and cut around the pencil line to create a template.
6. Trace around the template onto the sheet of black paper.
7. Cut around the drawing carefully.
8. Glue the silhouette to the white paper. Frame it and hang it on your wall.

Make Hand Shadows

All you need is a light source and your hands

Conjure up a menagerie of animals and a crazy cast of people to star in your show. Simply use your fingers, palms and arms to make shapes and hold those outlines between your light source and a surface (your 'screen'). A good torch, a lantern, a candle or even a streetlight will all work well. Try casting your shadows on walls, on the floor, on a sheet hung on a line, or the inside of a tent.

Experiment with distances – make the shadows larger by moving closer to your light source, hold your torch down low or up high and see how it affects the look of your shadows. Now, let your imagination go wild and create your own shadow stories.

Use both hands to come up with more complex characters.

Leaf Lanterns

Nature-inspired lights for dark nights

Many festivals use light as a metaphor for life, hope and rebirth – Germany's St Martin's, Diwali, Hanukkah, Candlemas and Imbolc are all celebrated with lanterns and flames. These simple little lights would work for any of these events and showcase the beauty of the natural world.

YOU WILL NEED

- Greaseproof paper
- Small round shallow container
- Card
- Leaves *(dry them out if wet)*
- PVA glue
- Brush or glue spreader
- Tea light *(battery-powered for safety)*
- String *(optional)*

1. Decide how tall you would like your lantern to be and cut the greaseproof paper accordingly. The length should be a little longer than the circumference of the container.
2. Stick leaves to the greaseproof paper using glue, spreading them out so they don't overlap.
3. Stick another layer of greaseproof paper on top and use books or heavy objects to flatten the leaf-covered paper.
4. Stick the bottom side of the paper into the base of the container, curving it to fit neatly. Alternatively, make a hoop out of card, and stick the paper into that hoop.
5. Make a hoop of card to stick around the top of the lantern. Turn on the tea light and put it inside the lantern.
6. Make a handle from string if you'd like to carry your lantern.

Pin-Prick Art

Make pictures that come alive when you shine a light

Pin-prick art is easy for little fingers to create, but still fascinating and complex enough for teenagers, making this a scalable, fun activity for families to try together. This form of craft was a popular pastime among the upper and middle classes in 18th century Britain and America. Special shops sprang up to supply amateur art enthusiasts with fine papers in all thicknesses and colours of the rainbow.

YOU WILL NEED

- Paper – *coloured, black or white.*
- A soft pencil
- A thick piece of cardboard or foam
- Push pins, sticky tape or adhesive putty
- A small sharp nail, tack, awl or needle
- A light source – *table lamp, torch, LED tea light or candle*

1. Draw a design in pencil on the reverse of your paper.
2. Using pins, tape or sticky putty, fix your paper to the cardboard.
3. Following the pencil lines, carefully use the sharp object to prick through the paper. Be careful not to tear the paper. The popping sound you get when you push through is very satisfying!
4. Older children might like to fill in areas of the paper with pin pricks to create different effects. Lots of dots close together will let more light through and that area will be brighter.
5. Hold your picture in front of your light source and watch it spring to life.
6. Experiment with using different objects to make holes and create new textures and patterns.

Why Not...

Create a display in your front window, leave the curtains open and hold an art exhibition for passers-by.

Or use your art to make beautiful lanterns – simply tape your masterpiece into a cylinder shape, and stand it around an LED tea light.

Sharp Safety!

- **Young children should be supervised when using sharp objects.**

Learning After Dark

Use Morse Code to Communicate

Send messages secretly and discreetly

Although most people have a mobile phone, and walkie talkies are relatively cheap, it's fun to learn to send messages to your friends over distances in the dark using only a torch or whistle. After all, you never know when your phone might run out of charge.

YOU WILL NEED

- At least two people
- A torch or whistle
- Pen and paper
- A print-out or copy of the Morse code chart
- Card to cover your torch *(optional)*

1. Signalling with Morse code can be trickier than it looks, so initially it's best to practise at short range. Each letter of the alphabet is represented by a series of dots (short flashes) and dashes (longer flashes), and you'll use a torch or whistle to create these dots or dashes. Some people find covering the torch with card a little easier than turning it on and off rapidly.
2. If you think of the code in beats, a dot is one beat, a dash is three. The space between each dot or dash is one beat. The space between letters is three beats. The space between words is seven. The receiver should write down each letter as they receive them, and your message will start to emerge. Begin with small words and build up to longer messages over bigger distances. Soon you'll be able to flash messages over streets, valleys or campsites.
3. Use your skill creatively. Use your bedroom light to transmit messages to a friend outside, whistle to each other while camping at night, or even use hand squeezes or table tapping to talk secretly to a friend or relative when others are about.

Dots and Dashes

A •–	B –•••	C –•–•	D –••	E •	F ••–•
G ––•	H ••••	I ••	J •–––	K –•–	L •–••
M ––	N –•	O –––	P •––•	Q ––•–	R •–•
S •••	T –	U ••–	V •••–	W •––	X –••–
Y –•––	Z ––••	1 •––––	2 ••–––	3 •••––	4 ••••–
5 •••••	6 –••••	7 ––•••	8 –––••	9 ––––•	0 –––––

TELL ME ABOUT: Morse Code

The Morse code communication system was invented by Samuel Morse, Joseph Henry and Alfred Vail, around 1837, alongside the electrical telegraph system. It uses a system of dots and dashes to represent letters. Each letter has its own pattern of these two symbols. Dots and dashes can be relayed in many different ways – visually, audibly or even via touch. By using torch flashes, whistles or knocks you can accurately relay messages across distances.

Helpful shortcuts

SOS or Help! ••• — •••

Hello —••• •••—

Stop *(or, "My message has ended")* •– •– •

Wait *(or, "I have more to say, but in a little while")* •– • • •

Over *(or, "OK, I'm done for now, it's your turn")* –•–

Understood • • •– •

Roger *(or, "I have received your message")* •–•

? ••–••

Yes – •– • *(C – as in 'si' – yes in Spanish)*

No –• *(N)*

Message session ending *(or "I'm off")* –•–• •–••

Get Super-Powered Night Vision

Train yourself to see in the dark

Want pumped-up perception? Get your night vision on and you'll be able to see in the dark. And the good news? It doesn't involve having to eat a ton of carrots. Humans are diurnal creatures – we mainly go about our business in the day – so we can see best when it's light. However, the human body is a remarkable machine, and, given time – about 45 minutes or so – our ability to see in the dark can improve.

Good Pupils

First, we need to understand how our pupils work. Pupils are the little holes in the centre of the eye that open and close to let in the light. The bigger they are, the more light they let in. Try a simple experiment to see how yours change and adapt to different lighting conditions.

YOU WILL NEED

- Torch
- Mirror
- Magnifying glass *(optional)*

1. Stand in a well-lit area.
2. Look into the mirror and look at the black dot in the centre of your eye. Use a magnifying glass to help if you have one.
3. Make a note of how large your pupil (the black dot) is.
4. Find a darker place to stand and stay there for a few minutes.
5. Using a torch, look again in the mirror at your pupil. Be quick! Has your pupil got bigger? Or smaller? Or has it stayed the same?
6. Try the experiment in different lights.

Power Up

Want even better night vision? Try this method of turbo-charging your sight in the dark. The back of your eyes are covered in cells called photoreceptors. These cells are sensitive to light and come in two flavours; rods and cones. The rods are the cells that help us to see better at night, as they're more sensitive to light and movement. Try this experiment to flex your rods and get that night vision pumped!

YOU WILL NEED

- Your eyes
- Torch with a red-light option or covered with red plastic wrap or cellophane

OPTIONAL

- Watch
- Something waterproof and comfy to sit on. *You'll be waiting around for a while!*
- Snack

1. Find your spot. You'll need somewhere really dark, as far away from streetlights, lit windows, fires or lamps as possible. If you're in a garden, move around until you're in the darkest area. If you're out and about, get as far away from lit roads as you can.
2. Turn all non-red torches, mobile phones or other light sources off. If you want to look at your watch, use the red light torch.
3. Now wait. And wait. After about 15 minutes your pupils will be at their widest. But there's a long way to go.
4. Wait some more. And some more. Perhaps try some nature listening (see page 82) to pass the time.
5. After 45 minutes, your rods will have absorbed enough rhodopsin (a protein found in your eye that aids night vision) to help you see movement and shapes in the dark. Peer about using your new powers. What can you see? Can you see colours?
6. Try looking at things out of the corner of your eye. A star, someone's face. Can you see it better? If so, it's thanks to those rods again – they're mostly found around the edge of your retina, in your peripheral vision.
7. Don't use a torch or look at a light source, or you will reset your rods, and you'll have to start the process again (ugh!).
8. Now, go and put your night vision to good use. Of course, your super-powers will only last as long as you don't step into the light. But hey, every hero has their Kryptonite!

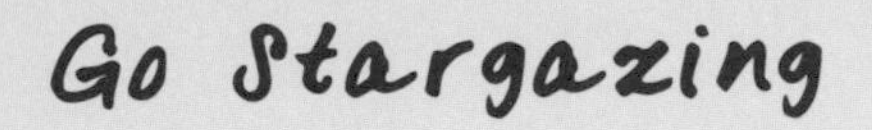

Go Stargazing

Look up and unlock the secrets of the universe

Star-spotting is a year-round activity:

Winter nights are perfect for exploring the sky; on really short days the first stars or planets might even be emerging on the walk home from school.

In summer you can stay out late giggling on a blanket under the night sky, staring deeply into space, spotting distant milky clusters of stars, creating your own constellation stories or even, if you're lucky, seeing a meteor zooming through the sky.

Discovering that a glowing light in the sky is a planet can be a magical experience for a child. Buy a star guide or print a star map to help you itentify what's what.

Stargazing requires nothing more than your eyes and a clear night. Follow these tips to get started and turn this activity into a regular family event.

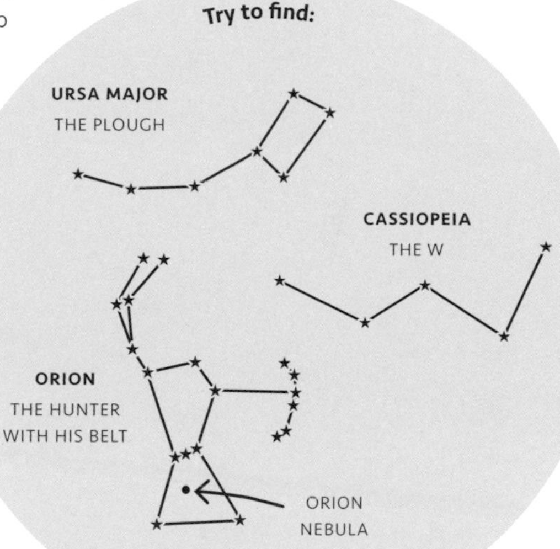

Be comfortable! Pack warm clothes and some sweets or a flask of hot chocolate for a treat. Take something warm and waterproof to lie or sit on, because even on a dry night there will be dew and damp.

Turn your house lights out to let your eyes get used to the dark – it might take a while for them to adjust. Try moving around to block out as much extraneous light as possible – you might only have to shift a metre to the left to let a house or tree block out a street light. Use a red torch, or put some red cellophane over a white one to ensure your eyes retain their night vision (see page 124).

Head out of town, away from any light pollution. Venture into the countryside for the darkest skies; mountains provide great vantage points, as do cliffs or wide plains. Look online at lightpollutionmap.info to find light-pollution-free places.

Learn a few constellations around the ones you've already spotted. Look for Orion's hound; Sirius, the Dog Star, who's hunting Taurus the bull, just ahead of him.

Although you can get great starchart apps, it's best to avoid looking at phone screens while you're out, as it'll mess up your night vision and can be distracting.

Heavenly Hints

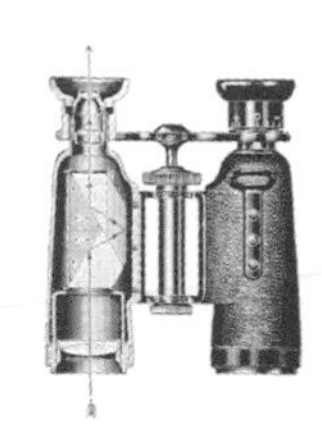

Pack binoculars – although not essential, binoculars will give a more panoramic view of the skies than telescopes, putting stars into context.

Stargaze before the moon is full, and the nights are darker. Look up the next full moon dates before you plan a special trip and avoid them.

Try spotting planets – when Venus is out, it's usually the brightest object in the sky; the first thing you see in the evening, or the last to disappear in the morning.

Look at the Moon. A close-up view will reveal craters, lava plains, and dark spots called seas or 'maria'. See if you can spot one of the biggest craters, Tycho, which is situated near the bottom of the Moon and has distinctive white 'rays' that fan out from its edges.

Most importantly, always be led by your imagination. It's not necessary to stick to the rulebook. Why not find your own patterns in the sky, name them, and create wild stories about how they got there?

Shooting Stars

Seeing a real-life meteor streak across the sky is a magical experience. They're created when pieces of asteroids or comets slam into the Earth's upper atmosphere and burn up due to friction against air particles. Their trail can glow from anything from a second to a few minutes. Get up really early in the morning for a grandstand view. Lie back on a blanket, let your eyes adjust to the dark, and be patient.*

You can spot shooting stars at any time of year, but there are certain periods when the skies are much busier, called meteor showers.

The following meteor showers are visible in the northern hemisphere, some in the southern too:

- **Quadrantids** December/January
- **Lyrids** April
- **Perseids** August
- **Orionids** October
- **Leonids** November
- **Geminids** December

* Don't forget to wish on each one you see!

LED Throwies

Simple, tiny lights for fun and games

These little self-powered lights stick to ferro-magnetic surfaces and look spectacular in groups. Make up a bunch of them and use them for all kinds of arty fun and mischief!

Magnetic Throwie

For each Throwie:

YOU WILL NEED

- ✗ Two-pin light emitting diode
  (any size or colour – find them online)
- ✗ 3v lithium battery CR2032
  (the little round ones)
- ✗ Small magnet
- ✗ Duct tape

1. Your LED should have a long and short leg. Slide the longer over the positive (+) side of the battery.
2. Pinch the LED – it should start to glow!
3. Wrap a thin slice of tape around the battery and LED two times.
4. Place a magnet on the positive side and wind another piece of tape around that too.
5. You're done! Now make some more. Each throwie should last for at least a week. Include a switch to give them a longer life.

Use your magnetic throwies to...

- Create patterns on a metal board
- Make words on metal surfaces in your neighbourhood
- Light up your fridge door

Non-Magnetic Throwie

Make as described (left) but without the magnet. Use them to...

- Add to your artworks to make them come alive at night
- Sew into your clothes
- Make light-up greetings cards
- Make light-up badges
- Put in masks to make their eyes glow
- Power lanterns
- Light up slime
- Attach to paper aeroplanes
- Bring your LEGO creations to life
- Make headlights for cardboard cars

How Throwies Work

By attaching the LED legs to the battery, you are creating a circuit. This unbroken loop allows electricity from the battery to pass through the LED light, and make it glow.

Make Sparks

Discover the science behind these tiny flashes of light

In the dark, you can see sparks fly. Creating tiny thunderstorms using what you have to hand is a magical experience. Here are two different methods to create these little flashes, one harnesses the power of electricity, the other uses heat.

Flints

YOU WILL NEED

- Flint rocks
- Piece of high carbon steel (stainless won't work). *Steel files (from a hardware store) will work well*
- Safety goggles

- Flints are forms of hard rock such as quartz, agate, chert and jasper. You can find them on beaches, in chalk cliffs, and in fields. Their insides have a very smooth, almost glassy appearance. If you split a stone, then hit it against another in the dark, you'll see sparks fly. However, it's safer and more effective to use a steel.
- Drag your steel across the surface of the flint, or strike the stone in a downward motion, and you'll see bright white lights start to fly. Remember to wear your safety goggles.

Heat Power

When you strike your steel on a flint, tiny, tiny pieces of iron are shaved off, which exposes fresh metal. As these fragments fly through the air, they make contact with oxygen and spontaneously ignite, giving off heat and light as they oxidise.

Static Showtime

YOU WILL NEED

- Clean, dry socks
- Carpet
- Balloon

- The easiest way to light up a room (well, sort of) is to use your feet. Shuffle along a carpet, then touch something metal or a friend. Watch those sparks fly!
- Blow up your balloon and rub it on your hair, a soft jumper, or ball of wool. Hold it near a metal object and see the tiny shards of light that fly between the two objects.

How it Works

When we rub objects, we are charging them with electricity. Electrons, which have a negative charge, can be transferred – by rubbing – from one object to another, leaving one object positively charged, the other negatively. If an object is made of insulating materials, it 'holds' its new charge until brought near to something that conducts well – metal or a human standing on the ground. As the two objects get closer, those electrons will leap across to the conductor, and you'll see a spark.

Spark Safety!

- **Don't make sparks near any dry grass or kindling – they will cause a fire.**
- **Be very, very careful if you're banging flints together – keep away from other people and protect your eyes with safety goggles.**
- **Don't touch electronics if you're charged with static electricity – you could damage them.**

Compare Senses
Explore your sensations
with fun games

Our five senses are touch, smell, hearing, taste and sight. In the dark with a lack of visibility, your four other senses compensate; you can hear in more detail, smells seem stronger, your taste buds pop and your fingers feel more sensitively. Try experimenting to see how your senses work differently in the dark.

Taste

Blind dining, where restaurants serve dishes in the dark, has become a popular way to eat out. Why not try your own take on it at home? It's a simple concept but requires some preparation and thought.

This activity is best for older children and teens, but it can be adapted for younger children, although you may need to feed them.

YOU WILL NEED

- Blindfolds *(scarves will do)*
- Cutlery and unbreakable plates
- Unbreakable glasses or cups and straws
- Food

1. Plan your menu. You could serve a whole dinner, part of a dinner, or a selection of small samples. Your choice should not be too messy and, if it needs cutting, you should do that in advance. Try to avoid dishes that are easily identifiable by touch. Keep a little of each dish back.
2. Prepare your feast, making sure that your diners are well out of the way. Put your creations onto a plates or into bowls.
3. Add straws to drinks to avoid spillage.
4. Put the ready-plated food and drinks onto a table – make sure your guests don't see their dinner.
5. Blindfold your guests and lead them to the table. Give them a knife and fork each and show them where their plates are.
6. Turn out the lights. Your guests can now remove their blindfolds.
7. Let them eat and listen in to what they think they're scoffing.
8. When they've finished, turn on the lights and bring in the samples of the menu you kept back.
9. Compare notes – what did they identify? What dishes left them stumped? Why do you think that was?

Smell

This game is good for older kids, although younger children will enjoy the sensations that accompany it.

YOU WILL NEED

- Series of small containers
- Dry marker
- Notepad for each participant
- Small torch for each participant
- Smelly (and not quite so smelly) things to sniff, such as: *coffee, tea, cinnamon, lemons, garlic, mint, vinegar, plain soap, dirt, tomatoes, cucumbers, onions, ginger, cloves,*
- If you want to use liquids such as rainwater or perfume, drip them onto a ball of cotton wool.

1. Put a little of each smelly material into its own individual pot. Number the pot using marker pen and make a note of what each pot contains.
2. Ask the participants to sit in a circle or around a table. Turn off the lights.
3. Pass each pot around the table or circle. Everyone should stay silent during the process. No distractions, no hints.
4. Ask each participant to inhale deeply from one of the pots and try to guess its contents.
5. After the pot has been passed around the whole group, the participants should switch on their torches and write down the number of the pot, and what they think it might contain.
6. After all the pots have been passed around, the lights should be switched on, and lists marked.
7. The champion smeller is the person with most correct answers.

Sniffing Around

- **Your ability to smell is thanks to a little patch of tissue high up your nose – your olfactory neurons – a group of sensory cells connected directly to your brain. It's also linked to thousands of nerve endings on the moist areas of your head – the eyes, nose, mouth and throat.**
- **A hypersensitive sense of smell is called hyperosmia – some people with the condition can even get ill after sniffing certain smells.**
- **Colds, dental issues and obstructions can all affect our ability to identify a whiff, as can smoking, head injuries, or some medical treatments.**
- **Our sense of smell becomes weaker as we grow older – how did the younger participants compare to the older in your test?**

Touch

It can be surprisingly difficult to identify objects using only your sense of touch. This simple experiment/ game demonstrates how nerve endings in our fingers send signals to our brains – signals that can be confusing! This game is for small kids, although older ones will enjoy it too.

1. Take a few familiar and unfamiliar object and place them into a bag. Try toy cars, favourite teddies, plasticine, wool, feathers, acorns, sticks and pebbles.
2. Turn the lights out.
3. Pull each object out of the bag in turn and pass it around the players (use a small torch to see what you're doing if needed). Explore each item with your fingers in order to identify it. Don't call out what you think the object is – wait until everyone has had a chance to feel it.
4. On the count of three, all participants should shout out what they think they've just been touching.
5. Give each player a point for a correct guess. The winner is the player with the most points at the end of the game.

Hearing

When we listen in the dark, we hear more. More of an exercise or meditation than a game, this way of listening to music brings it alive.

YOU WILL NEED

- Something comfy to lie on
- A music source

1. Lay on something comfortable and put on your music – classical pieces and slower dance tracks with no lyrics work especially well – turn off the lights and listen hard.
2. Try not to talk to each other, just lay back and take in the music. Listen hard – identify each instrument and how it fits with the others, take in the lyrics, or find the pulse, the beat of the song. Or simply let it wash over you completely and let it slide into your ears and your brain.
3. This method of listening focuses your attention fully on the music, and can almost be overwhelming. It's a good activity to do before bedtime – try gentle, pastoral pieces for maximum relaxation.

Spot the International Space Station

Watching this spacecraft track across the night sky can be strangely emotional

This tiny, bright dot is a satellite that's home to real people; scientists who live and work on the research ship that constantly orbits the Earth. These people spend their lives observing weather patterns, performing experiments in the microgravity laboratory, and stake a human presence in space. Seeing it tracking far above our heads gives us a sense of perspective and respect for the universe and science.

This satellite flies at a relatively low height – approximately 400 km (250 miles) above the earth – and reflects the light of the sun, making it the third brightest object in the night sky. Which means it's easy to spot! However, there are a few factors you have to consider if you want to see the ISS:

- The sky must be dark or darkening – any time between dusk and dawn.
- The night must be clear – you won't see the ISS if it's cloudy.
- The ISS must be orbiting over your location.

These factors may seem tricky to coincide, but help is on hand. Nasa's online ISS tracker can tell you when you'll spot the satellite from your location.

Search online at spotthestation.nasa.gov, enter your nearest town, and you'll find out when the ship will be above your heads. You can even sign up for email or text alerts that give you a ping ahead of visible flights. It helps to be prepared because the window of viewing time for the ISS can be very short – always under ten minutes, and often less than five. The site will also tell you which direction to look and the height the station will be above the horizon in degrees.

A little preparation will result in the most spectacular viewings; get yourself somewhere you can see a lot of sky – an open space with few trees or a high vantage point.

- Take a compass so you know in which direction to look.
- Wrap up warm and take something to sit on.
- Get to your viewing place in plenty of time, turn off your light sources, and let your night vision kick in (see page 124).

You don't have to go to any trouble to spot the satellite – you can get just as thrilling a show walking home from school on a dark winter evening – but a plan will make it a little more special, and smaller viewers will be more likely to be able to spot it easily.

The Space Station moves in a distinctive straight line. It has no flashing lights and looks like a very bright star.

Watch it glide across the sky, imagine the scientists on board, and what they can see, and blow your minds.

Christmas Magic

Some people say that the ISS reminds them of Father Christmas riding his sleigh across the night sky. Use the NASA sight to look up December sightings and make the month a little more magical.

Space Station Facts

- The Station was launched in 1998 and is expected to continue orbiting until 2030.
- Occupants carry out experiments in physics, astronomy, meteorology and other scientific fields.
- It circles the Earth every 90 minutes.
- Its six laboratories and living spaces are powered by solar energy.
- It measures 108.5m (356ft) by 72.8m (239ft) – a little larger than a full-sized football pitch!
- It weighs 450 tons – roughly 450 times heavier than a car.

Index

About Kate Hodges

As Features Writer for The Green Parent and mother to nine-year-old twins, Kate is passionate about screen-free fun for families. She has over 25 years' experience in print journalism and is the author of *London in an Hour*, *Rural London*, *I Know a Woman*, and *Warriors, Witches, Women*. Kate has written for publications including *Just Seventeen*, *Smash Hits*, *The Guardian*, *Kerrang!* and *NME* and lives in East Sussex.

About the photographer

Jeff Pitcher started taking pictures in earnest after the birth of his second (of five) children. His work has appeared in *The Guardian*, *Shindig!*, *The Green Parent*, *Practical Photography*, *Black and White Photography*, *Amateur Photographer*, and he's held three major exhibitions. His photographs illustrated the best-selling guide to the capital's green spots, Rural London.

Picture credits

All photographs © Jeff Pitcher except for the folllowing:

17 Becart/Getty
80t Nordelch/Wikimedia Commons/CC BY
80b Stephen Orsillo/Alamy
81t Sreejithk2000/Wikimedia Commons/CC BY
81m Dr Henry Oakeley/CC BY
81b David Stang/Wikimedia Commons/CC BY
86t Wellcome Collection/CC BY
86b Wellcome Collection/CC BY
87t Wellcome Collection/CC BY
87b Wellcome Collection/CC BY
89 Ozza Okuonghae/CC BY
93l Wellcome Collection/CC BY
93r Wellcome Collection/CC BY
94 Arthur Dias, PT/CC BY
99 (illustration) IconMark, PH/CC BY
123r shemoto/iStock/Getty
128t Raymond/Wikimedia Commons/CC BY
128b Svdmolen/Wikimedia Commons/CC BY
129 Chumphon Whangchom/EyeEm/Getty
139 Xshadow/Wikimedia Commons/CC BY

Acknowledgements

Huge high-fives to all the children who took part:

The Hastings Hustlers
Samuel Bradnum, Edward Hynes, Emily Hynes, Arthur Jenkinson, Dusty Jenkinson, Vinnie Jensen, Dee Dee Yarde, Jackson Yarde

The Crawley Crew
Cora Brown, Ivy Brown, Molly Brown, Phoebe Fassam, Jude Mitchell

The Greenwich Gang
Naveah Marie Beadell-Alexander, Olivia Bolton, Zachariah Bolton, Ben Brown, Daniel Brown, Poppy Bushnell, Elizabeth Clarke-Williams, Nia-Larae Fernandez-Folks, Omar Gerber, Saffiya Gerber, Elisha Gles, Kaelah Gles, Kiana Gles, Kessiah Hart, Sophie Jessup, Ayana Kennedy, Jahzara Kennedy, Nyashanti Kennedy, Evie Knifton, Chinua Nwankwo, Chizram Nwankwo, Iris Page, Ailsa Pyne-Shaw, Maiá Williams

Grateful thanks to Fiona Machen Harrison at Christchurch School Community Garden in Greenwich (@growingforgold on Twitter) and Debbie and Jane at Waterlea Adventure Playground in Crawley.

Huge thanks to Juliet Pickering and Hattie Grünewald, wonder agents at Blake Friedmann. Emma Bastow for her super-chilled editing. Nicky, Liz, Isabel, Jennifer, Melissa and Jessica and the rest of the gang at Quarto.

Disclaimer
Although every reasonable effort has been made to ensure all the information in this book is correct as at the date of its publication, the author and publisher do not assume responsibility for and hereby exclude to the fullest extent possible at law, any and all liability for any loss, damage, injury, illness or loss of life caused by negligence (including incorrect information in this book) or mistakes in the interpretation of the information in this book. Reliance on the information in this book is at the sole risk of the reader.

Author's note on safety
The activities in this book are designed for the whole family. You know your children better than I do, so you will know the extent of their abilities and the level of supervision they need. Some will grab a book like this and be utterly sensible, others less so. It pays to never under- or overestimate a child's ability, but to understand the risks and manage them well. Have fun after dark but please follow these guidelines, which will help you to stay safe and look after your environment.

General guidelines

- Respect all wildlife and be considerate to other users of public places. Whatever you bring out into the wild, take it back home with you.
- Leave outdoor spaces as you find them.
- Only use nightlights or candles under adult supervision and never leave lanterns unattended.
- Only use a knife or sharp tools if you have been given permission and shown how to use them safely.
- Make sure everyone is aware of the potential dangers of using sharp tools; accidents usually happen when people are messing around.
- Always put tools away when not in use; never leave them lying around.

Fire safety guidelines

- Never make fire unless you have permission to do so and adults are around to supervise.
- Make fires well away from overhanging trees and buildings. Make fires on mineral soil, in a pit or (preferably) in a fire pan.
- Never light a fire in windy or very dry weather conditions.
- Never leave a fire unattended.
- Have a supply of water nearby to put out the fire or soothe burns.
- Use as little wood as you can and let the fire burn down to ash. Once it is cold, remove all traces of your fire.

First published in 2020 by White Lion Publishing,
an imprint of The Quarto Group.
The Old Brewery, 6 Blundell Street
London, N7 9BH,
United Kingdom
T (0)20 7700 6700
www.QuartoKnows.com

A catalogue record for this book is available from the British Library.

ISBN 978-0-7112-4622-5
Ebook ISBN 978-0-7112-4623-2

10 9 8 7 6 5 4 3 2 1

Design by Isabel Eeles

Printed in China